Population Health Intervention Research

Health geographers are well situated for undertaking population health intervention research (PHIR), and have an opportunity to be at the forefront of this emerging area of inquiry. However, in order to advance PHIR, the scientific community needs to be innovative with its methodologies, theories, and ability to think critically about population health issues. For example, using alternatives (e.g. community-based participatory research) to traditional study designs such as the randomized control trial, health geographers can contribute in important ways to understanding the complex relationships between population health (both intended and unintended consequences), interventions, and place. Representing a diverse array of health concerns ranging across chronic and infectious diseases, and research employing varied qualitative and quantitative methodologies, the contributions to this book illustrate how geographic concepts and approaches have informed the design and planning of intervention(s) and/or the evaluation of health impacts. For example, the authors argue that geographically targeting interventions to places of high-need and tailoring interventions to local place contexts are critically important for intervention success. Including an afterword by Professor Louise Potvin, this book will appeal to researchers interested in population and public/community health and epidemiology as well as health geography.

Daniel W. Harrington is an Epidemiologist Lead at Public Health Ontario, and Adjunct Assistant Professor in the School of Public Health and Health Systems at the University of Waterloo, Canada.

Sara McLafferty is Professor of Geography and Geographic Information Science at the University of Illinois at Urbana-Champaign, USA.

Susan J. Elliott is Professor in the Department of Geography and Environmental Management at the University of Waterloo, Canada.

Geographies of Health

Series Editors

Allison Williams, Associate Professor, School of Geography and Earth Sciences, McMaster University, Canada
Susan Elliott, Professor, Department of Geography and Environmental Management and School of Public Health and Health Systems, University of Waterloo, Canada

There is growing interest in the geographies of health and a continued interest in what has more traditionally been labeled medical geography. The traditional focus of 'medical geography' on areas such as disease ecology, health service provision, and disease mapping (all of which continue to reflect a mainly quantitative approach to inquiry) has evolved into a focus on a broader, theoretically informed epistemology of health geographies in an expanded international reach. As a result, we now find this subdiscipline characterized by a strongly theoretically informed research agenda, embracing a range of methods (quantitative, qualitative, and the integration of the two) of inquiry concerned with questions of: risk; representation and meaning; inequality and power; and culture and difference, among others. Health mapping and modeling has simultaneously been strengthened by the technical advances made in multilevel modeling, advanced spatial analytic methods, and GIS, while further engaging in questions related to health inequalities, population health, and environmental degradation.

This series publishes superior quality research monographs and edited collections representing contemporary applications in the field; this encompasses original research as well as advances in methods, techniques, and theories. The *Geographies of Health* series will capture the interest of a broad body of scholars, within the social sciences, the health sciences, and beyond.

Population Health Intervention Research

Geographical perspectives

Edited by Daniel W. Harrington, Sara McLafferty, and Susan J. Elliott

LONDON AND NEW YORK

First published 2017
by Routledge

2 Park Square, Milton Park, Abingdon, Oxfordshire OX14 4RN

52 Vanderbilt Avenue, New York, NY 10017

Routledge is an imprint of the Taylor & Francis Group, an informa business

First issued in paperback 2020

British Library Cataloguing in Publication Data
A catalogue record for this book is available from the British Library

Library of Congress Cataloging in Publication Data
Harrington, Daniel W., editor. | McLafferty, Sara, 1951- editor. | Elliott, Susan J., editor.
Title: Population health intervention research : geographical perspectives / edited by Daniel W. Harrington, Sara McLafferty and Susan J. Elliott.
Description: Milton Park, Abingdon, Oxon ; New York, NY : Routledge, 2017. | Includes bibliographical references and index.
Identifiers: LCCN 2016009867| ISBN 9781472457257 (hbk) | ISBN 9781315601502 (ebk)
Subjects: LCSH: Medical geography.
Classification: LCC RA791 .P67 2017 | DDC 614.4/2–dc23
LC record available at https://lccn.loc.gov/2016009867

ISBN: 978-1-4724-5725-7 (hbk)
ISBN: 978-0-367-66819-8 (pbk)

Typeset in Times New Roman
by Cenveo Publisher Services

On behalf of his co-Editors, Dan Harrington wishes to acknowledge the efforts of each of the contributing authors. Thank you for your seemingly endless patience. He would also like to dedicate this book to Ophelia Rose, who was born and celebrated her first birthday as this book was taking shape. Dad's dance card just opened up a little.

Contents

Figures

Tables

Contributors

Kirsten Beyer is Assistant Professor of Epidemiology in the Institute for Health and Society at the Medical College of Wisconsin and Adjunct Assistant Professor of Geography at the University of Wisconsin-Milwaukee. Trained in geography and public health, Dr Beyer's work focuses on identifying spatial patterns of disease and injury, understanding complex human–environment processes that create those patterns, and developing and evaluating community-based interventions informed by a social-ecological understanding of risk. She teaches in the PhD Program in Public and Community Health at the Medical College of Wisconsin.

Sandra Bogar is a doctoral candidate in the Public and Community Health program at the Medical College of Wisconsin. Her research interests broadly include health inequities related to social and environmental determinants of health, and research approaches which maximize community involvement and are driven by community priorities.

Brian E. Cook is a Health Research Specialist at Toronto Public Health. As research lead for the Toronto Food Strategy, he works with a broad range of partners to implement unique or underdeveloped solutions to expand access to healthy and affordable food. In 2015, he completed a federally funded evaluation of "healthy corner stores" and a mobile fresh produce market in lower-income Toronto neighborhoods. The research tested strategies to support convenience stores to sell healthy, affordable food in profitable ways, and assessed the Mobile Good Food Market, a retrofitted City of Toronto bus that sells fresh produce. Alongside project implementation, Brian works closely with staff from other city divisions and agencies to develop municipal policy that enables the scaling up of creative food system initiatives.

Sarah Dickin is a Research Fellow at the Stockholm Environment Institute and has an inter-disciplinary background with a PhD in Geography and a BSc in Environmental Science from McMaster University, Canada. Her research interests are centered on understanding how global environmental change is impacting health and well-being, particularly at the interface of water and health challenges, using a range of qualitative, quantitative, and GIS approaches. Prior to joining SEI, Sarah was a researcher at the United Nations University Institute for Water,

Environment and Health in the Water and Human Development program and worked on a range of water and health topics, such as how national aspirations for improving water and sanitation correspond to Sustainable Development Goal targets. As a postdoctoral researcher she leads work examining climate change uncertainty among stakeholders in the Great Lakes with the Ontario Climate Consortium and Environment Canada, and assessed climate change health impacts in the MENA region for the UNESCWA RICCAR initiative.

Susan J. Elliott is a health geographer in the Department of Geography and Environmental Management at the University of Waterloo, Canada. She has authored over 150 peer-reviewed publications in the area of environment and health. Much of her recent environment and health research focuses on science-policy bridging and the knowledge-to-action link.

Michael Emch is Professor and Chair of Geography at the University of North Carolina at Chapel Hill. He is also Professor of Epidemiology and a Fellow of the Carolina Population Center. He has published widely in the subfield of disease ecology, mostly on infectious diseases of the tropical world. He directs the Spatial Health Research Group (http://www.cpc.unc.edu/projects/spatialhealthgroup).

Daniel Fuller has an MSc in Kinesiology from the University of Saskatchewan and a PhD in Public Health from the Université de Montréal. At the time of writing his chapter with Erin Hobin he was an Assistant Professor in the School of Public Health at the University of Saskatchewan. Starting August 1, 2016, he will be a Canada Research Chair in Population Physical Activity at the Memorial University of Newfoundland. His research interests include physical activity epidemiology, mobile sensing, active transportation, natural experiments, and social inequalities. Daniel is passionate about creating environments that promote health and reduce social inequalities. He rides his bike to work most days.

Daniel W. Harrington is an Epidemiologist Lead at Public Health Ontario, and Adjunct Assistant Professor in the School of Public Health and Health Systems at the University of Waterloo, Canada. As a quantitative health geographer, his research to date has focused on inequalities in access to primary and specialist healthcare in Ontario, individual- and neighborhood-level determinants of overweight and obesity, and perception of health risks. He is currently leading an outcomes evaluation for a large community-based program designed to address childhood overweight and obesity in Ontario, Canada.

Stephanie Heald has a Master's degree in Applied Geography with a focus on medical geography from the University of North Texas. She is currently completing a PhD in Geography at Oklahoma State University, where she is teaching cultural geography and researching the spatial patterns of mental illness in Oklahoma. Stephanie also has a Bachelor's degree in social studies education from Southeastern Oklahoma State University.

Erin Hobin is Adjunct Professor at the University of Waterloo and the University of Toronto, and a Scientist in Heath Promotion, Chronic Disease, and Injury

Prevention at Public Health Ontario. Her research focuses on developing and evaluating population-level interventions for chronic disease prevention, specifically in the areas of healthy weights, healthy eating, and alcohol control. Her current research includes studies investigating the efficacy and effectiveness of the proposed changes to the Nutrition Facts table in Canada and menu labeling legislation in restaurants in Ontario, an evaluation of a province-wide physical education policy on students' physical activity behaviors, and the efficacy and effectiveness of alcohol warning labels on alcohol containers sold in liquor stores in Canada. In March 2014, Erin was awarded a CIHR Operating Grant to examine the impact of an on-shelf nutrition labeling system on the nutritional quality of consumer food purchases in supermarkets in Canada. In total, Erin has been awarded more than $1 million in grant funding and published dozens of peer-reviewed articles in national and international journals.

Kelly Hoormann is a Program Coordinator in the Division of Epidemiology, Institute for Health and Society at the Medical College of Wisconsin. She has worked on outdoor physical activity interventions in the Milwaukee area and community and children's health initiatives in rural Panama. Her research interests include the impacts of environmental education, outdoor recreation, and general nature experiences on community health disparities, both locally and globally.

Moriah Iverson is the Program Manager for Community Engagement and Research at the Medical College of Wisconsin and Froedtert Hospital. She has been involved in several creative place-making initiatives in the City of Milwaukee, including the installation of a mobile technology lab on a revitalized abandoned railroad corridor, the installation of swings in a vacant underpass, and the creation of community gardens in vacant lots. Ms Iverson's research focuses on community health and wellness resources for individuals with disabilities.

Mark Janko is a graduate student in Biostatistics and Geography at the University of North Carolina at Chapel Hill. He is primarily interested in implementing hierarchical spatial models in a Bayesian setting to understand disease transmission, with particular interest in integrating spatial and molecular techniques for the study of malaria and other infectious diseases of public health significance.

Scott T. Leatherdale is Associate Professor and CIHR-PHAC Chair in Applied Public Health Research in the School of Public Health and Health Systems at the University of Waterloo, Canada. His research activities focus on population surveillance, developing and evaluating population-level health interventions, and creating the infrastructure required to facilitate population studies in chronic disease prevention. His work is purposefully designed to actively engage program and policy stakeholders at multiple levels (regional, provincial, and national), and is designed to have an impact on improving the health of large segments of the Canadian youth population.

Nathaniel M. Lewis is Lecturer (Assistant Professor) in the Population, Health, and Well Being (PHeW) Research Group at the University of Southampton's

Department of Geography and Environment (UK). A graduate of Queen's University and a recent Canadian Institutes of Health Research Postdoctoral Fellow, his work is primarily concerned with relationships between place, mobility, and gay men's health, particularly mental and sexual health. His work can be found in journals such as *Health and Place, Gender, Place and Culture*, and *Annals of the Association of American Geographers*.

Sara McLafferty is Professor of Geography and Geographic Information Science at the University of Illinois at Urbana-Champaign. She obtained her BA from Barnard College and MA and PhD degrees in Geography from the University of Iowa. Her research investigates place-based disparities in health and access to health services and employment opportunities for women, immigrants, and racial/ethnic minorities in the United States. Dr McLafferty has also written about the use of GIS and spatial analysis methods in exploring inequalities in health and access to healthcare. Her books include *GIS and Public Health* (with Ellen Cromley), *A Companion to Health and Medical Geography* (with Tim Brown and Graham Moon), and *Geographies of Women's Health* (with Isabel Dyck and Nancy Lewis).

Catherine L. Mah leads the Food Policy Lab, a multidisciplinary program of research in the policy and practice of public health, with a focus on health-promoting innovations in the food system. Her work integrates research and action on environmental contexts for consumption, using public policy as a tool to enable diverse food system stakeholders to support population health. Her work is funded by the Canadian Institutes of Health Research, Health Canada, and the Leslie Harris Centre of Regional Policy and Development. She holds faculty appointments at the Memorial University and the University of Toronto, Canada.

Leia M. Minaker is a Scientist at the Propel Centre for Population Health Research at the University of Waterloo, Canada. Her program of research focuses on dietary intake and tobacco use in the population, and in particular population health intervention aimed at improving dietary intake and decreasing tobacco use. She is interested in retail food environment interventions that are effective at improving health and are economically feasible for food retailers, and has worked closely with research, policy, and practice actors to evaluate such interventions.

Jared Olson is a doctoral student in the Public and Community Health program at the Medical College of Wisconsin. Previously a high-school educator, his research interests include health outcomes and equity as they relate to land use, food systems, political ecology, and praxis.

Joseph R. Oppong is Professor of Geography and Associate Dean of the Robert Toulouse Graduate School at the University of North Texas. His research focuses on the spatial patterns of disease and health services. His previous research included HIV/AIDS in Texas and Africa and tuberculosis in Tarrant County, Texas. His latest research, funded by the National Science Foundation through the Coupled Human and Natural Systems program, deals with modeling the spatial patterns of Buruli ulcer in Ghana.

Louise Potvin is Professor at the Department of Social and Preventive Medicine, School of Public Health, Université de Montreal and Scientific Director of the Centre Léa-Roback sur les inégalités sociales de santé de Montréal. She holds the Canada Research Chair in Community Approaches and Health Inequalities. Her main research interests are Population Health Intervention Research and the role of social environments in the local production of health and health equity. In addition to having edited and co-edited eight books, she has published more than 250 peer-reviewed papers, book chapters, editorials, and comments. She is a Fellow of the Canadian Academy of Health Sciences and the Editor in Chief of the *Canadian Journal of Public Health*.

Barbara L. Riley is Executive Director of the Propel Centre for Population Health Impact, a pan-Canadian enterprise founded by the Canadian Cancer Society and the University of Waterloo, whose mandate is to accelerate and advance the science of population intervention and its application. Dr Riley is an engaged scholar with a focus on evaluating complex public health interventions, examining scaling up processes, accelerating knowledge mobilization, and building capacity for population health intervention research and health evaluation.

Kerry L. Robinson is Senior Manager with the Centre for Chronic Disease Prevention at the Public Health Agency of Canada. She has 18 years of experience working in public health applied research, policy, and practice at local, provincial, and national levels in Canada. She is PhD-trained in intervention research, knowledge exchange, and health policy, and is a co-author of 19 peer-reviewed articles.

Warankana Ruckthongsook is a PhD candidate in Environmental Science at the University of North Texas (UNT). She has been working as a research assistant in Medical Geography at UNT since 2010 and is part of the Health and Medical Geography Research Group. Her research interests lie in the areas of medical geography, geographical epidemiology, and environmental health using applications of remote sensing, GIS, and spatial analysis. Her previous research includes an investigation of the impact of land use and land cover change on the spatial distribution of Buruli ulcer in south-west Ghana and race/ethnic disparities in HIV/AIDS in Texas. Her dissertation research focuses on the improvement of disease mapping methods such as the kernel density estimation (KDE) method.

S. Martin Taylor is Professor Emeritus at the University of Victoria, Adjunct Professor in the School of Public Health and Health Systems at the University of Waterloo, and Adjunct Professor in the School of Geography and Earth Science at McMaster University. He chairs the board of the Propel Centre for Population Health Impact. He is the co-author of two books and over 100 peer-reviewed articles in the field of health geography.

Chetan Tiwari is Associate Professor in Geography at the University of North Texas. His research emphasizes applications of GIS and spatial analysis, particularly disease mapping methodologies that account for geographical differences in population structures and spatial associations between environmental risks and adverse health outcomes. Dr Tiwari has a PhD from the University of Iowa.

Michelle M. Vine is an Evaluation Specialist in Health Promotion, Chronic Disease and Injury Prevention at Public Health Ontario and Adjunct Assistant Professor in the School of Public Health and Health Systems at the University of Waterloo, Canada. She is a social geographer, with a focus on the environment and health, population health intervention research, health policy implementation, and qualitative research methods. For her doctoral research (McMaster University), Michelle used a mixed qualitative design to evaluate the school nutrition policy environment in Ontario, examining local-level factors shaping policy implementation. Her current research is focused on assessing nutrition policy implementation in Ontario using data from the COMPASS study, and after-school nutrition programs in First Nations communities. She is currently working on a policy brief to link policy, practice, and research in school nutrition, and is a member of the editorial advisory board of Health Promotion Practice.

Cameron D. Willis is a Scientist and Research Assistant Professor at the Propel Centre for Population Health Impact, University of Waterloo, Canada. Dr Willis is an engaged scholar, working with policy and practice colleagues in collecting, analyzing, and applying relevant and rigorous evidence to pressing population health issues, particularly the prevention of chronic diseases. Through his appointment at Propel and interchange with the Public Health Agency of Canada, he provides leadership for a program of research focused on inter-organizational collaboration, including investigations of inter-organizational partnership outcomes, fostering learning and improvement in multi-sectoral partnerships (involving public and private institutions), and synthesizing knowledge for supporting systems change.

1   Introduction

*Daniel W. Harrington, Sara McLafferty,
and Susan J. Elliott*

As a scientific subdiscipline, health geography is a dynamic and reflective field that frequently applies a critical lens to its own position in the academic landscape (Kearns and Moon, 2002). This has been particularly evident through the "cultural" and "spatial" turns that have occurred since the latter part of the twentieth century (for a full treatment see Kearns and Moon, 2002; Brown, McLafferty, and Moon, 2009; Luginaah, 2009). Indeed, many will already be familiar with the emergence of health geography from the spatial-analytic thrust of "traditional" medical geography. This has been complemented through the application of critical social theory to interrogations of the influences of places on health, using a range of quantitative and qualitative epistemologies (Andrews and Moon, 2005). Concurrently, there has been an expansion of thematic areas of research that span, but are not limited to: population health, social inequities, disability and ageing, healthcare, housing, environmental risk and exposure, and chronic and infectious disease (Luginaah, 2009). The diverse approaches available to geographers offer unique perspectives for critically analyzing how spaces and places can shape the ways in which health is experienced by, and shaped by, individuals or groups of individuals. The point of this volume is to explicitly highlight the contributions that geographers can make and have been making to an important and emerging field of scientific inquiry: population health intervention research.

Studies published over decades of work from health geography and across other, related disciplines (e.g. epidemiology, public health, health promotion, community health) have shown important variations and inequities in health, health behaviors, and social determinants of health between and within populations and geographies. Importantly, geographers have advanced the understanding of how variations in health are influenced by context, and how health is experienced in the various socio-cultural, physical, economic, and political environments in which people live. While descriptive and analytic studies have produced crucial knowledge about the relationships between the social determinants of health and population health, more recently there has been an increased emphasis on studying population health interventions (Hawe and Potvin, 2009). We echo the arguments of others that geographers, and those from other disciplines, should assign equal effort to using evidence to try to narrow inequalities and inequities in health (Dunn, 2014). Certainly, this direction aligns with the population health approach (Rose, 1985; Public Health Agency of Canada, 2012) that underpins much of the empirical research in health geography and other closely aligned fields.

Population health interventions are policies or programs that act upon population health (directly or indirectly) towards altering an established course, be it absolute health burden, inequality, or inequity (Hawe and Potvin, 2009). A classic example from health promotion would be the introduction of the three-point seat belt in the late 1950s. Fitting automobiles with seat belts could be considered a program, as would social marketing efforts to increase population awareness of the benefits of their use. Whereas legislated mandatory seat-belt use and the enforcement of seat-belt laws would be considered policy interventions, together these interrelated policies and programs have proven to be impactful, cost-effective interventions that have saved over one million lives, globally (World Health Organization, 2009). This example clearly illustrates the concept of shifting the distribution of injury and automobile-related mortality by reducing exposure to risk at a population level (Rose, 1985).

The impacts of other population health interventions may not be as clear, depending on the complexity of the intervention. That is, if an intervention system has multiple components or targets multiple aspects of health, it may be difficult to identify or attribute any effects to one particular part (Craig *et al.*, 2008). Further, an intervention may have differential impacts on different segments of the population or unintended consequences. As such, the traditional gold standard for evaluating clinical interventions through randomized controlled trials may not be feasible or even appropriate on a population level (Hawe *et al.*, 2004; Glasgow *et al.*, 1999). Rather, it is often necessary to approach population health interventions with a diversity of methods and disciplinary knowledge. In this way, it is important to gain a deeper understanding of not only what interventions work (Frank, 2012), but also – to paraphrase Macintyre *et al.*, (2002) – what interventions work *for which populations* and *in what places.* It is likely unsurprising that this latter point is where we feel health geographers can make a significant contribution.

The geography of population health intervention research

With an understanding of what constitutes a population health intervention, we take the opportunity here to describe population health intervention research (PHIR), and explicitly identify a place for geographers in PHIR as a scientific endeavor. PHIR has been defined as:

> Research that involves the use of scientific methods to produce knowledge about policy and program interventions that operate within or outside of the health sector and have the potential to impact health at the population level. (Canadian Institute for Health Information, 2014)

These may include policies or program interventions that are not primarily designed to change health or health behaviors in the general population, but nonetheless have the potential to impact population health. For example, increasing access to public transit as a mechanism for easing traffic gridlock and reducing commuting times may concurrently increase active transportation (e.g. walking or cycling to transit stations) that could contribute to increasing daily physical activity and reducing stress, both important determinants of health outcomes such as cardiovascular disease and excess body weight.

If we understand that PHIR uses scientific methods to understand policies and programs that have the potential to impact population health, it follows that these generally operate by modifying upstream conditions that determine population health (Rose, 1985). It is somewhat interesting, then, that the PHIR definition (perhaps unintentionally) appears to privilege policies and programs related to the health sector by drawing attention to these through a specific reference. In practice, however, interventions that are effective in 'shifting the curve' by addressing the social determinants of health may be more aptly based in sectors other than health such as employment, education, social support services, healthy child development, transportation, or the built environment (Frankish, 2012; Public Health Agency of Canada, 2015). Then, if population health interventions encompass changes in these sectors, the influence of the social, economic, physical, or political environments in which health is shaped is quite clear. This provides a cozy place where geographers may see themselves contributing to PHIR and simultaneously provides justification for the efforts of all who contributed to this volume.

What type of knowledge is needed to understand population health interventions? Hawe and Potvin state "all systematic inquiry and learning from observing an intervention's process or implementation, impact or outcome is encompassed in 'intervention research'" (2009: I-9). Echoing this, Frankish explains "PHIR aims to capture the value and differential effects of interventions, the processes by which they create change, and the contexts within which they work best" (2012: S3). Both definitions deconstruct population health interventions into two discernible and equally important components that are targets for research. The first is specifically concerned with changes in health outcomes or determinants of health at the population level that can be attributed to the presence or intensity of an intervention. An outcomes evaluation is generally concerned with answering questions of intervention efficacy: *did the intervention work?* The second refers to the processes by which the intervention impacts on population health. According to Glasgow and colleagues' RE-AIM framework (1999), comprehensive evaluations should include assessment of five dimensions in order to understand why (or why not) an intervention impacts population health: reach, efficacy, adoption, implementation, and maintenance (Table 1.1). While outcomes evaluations are generally focused on reach and efficacy, other dimensions of an intervention can be unpacked through a process evaluation. Settings are central in health promotion theory (Green *et al.*, 2000), and assessment of adoption, implementation, and maintenance in particular are key to understanding the settings in which interventions are executed. Ultimately, both outcomes and process evaluations are necessary for fully understanding if interventions work, how interventions work, whom they work for, and where they work best.

Table 1.1 The RE-AIM framework

Dimension	Definition
Reach	Proportion of the target population participating in or exposed to the intervention
Efficacy	Success rate: changes in population health
Adoption	Proportion of settings adopting the intervention
Implementation	Fidelity: the extent to which the intervention was implemented as planned
Maintenance	The extent to which the intervention is sustained over time

Adapted from Glasgow *et al.* (2009).

This volume

The unofficial grandfather of medical geography, John Snow, conducted a classic example of PHIR in the mid-1800s. More than a century later, geographers are well situated for advancing population health intervention research, and we argue that geographers have an opportunity to be at the forefront of this emerging area of inquiry. In order to advance PHIR, the scientific community will need to be innovative with its methodologies, theories, and ability to think critically about population health issues. In particular, geographers can contribute in important ways to understanding the complex relationships between population health, interventions, and place.

Some have already taken up this charge, and this volume is meant to highlight that work and accentuate the inherently geographical nature of PHIR. We have solicited contributions representing a diverse array of methodologies applied to a broad range of health issues, from both emerging researchers in health geography (and closely related disciplines), as well as established scholars. Most are university-affiliated, though several are established scientists working in government public health organizations. All are well connected to the world of science-policy bridging. The contributors are primarily from North America, though efforts were made to ensure the inclusion of contributors who conduct PHIR in different contexts and communities around the world.

In organizing this volume, we have attempted to highlight topical and methodological connections between chapters, as well as broader themes linking geographical perspectives and PHIR. With respect to broad themes, chapters in the first half of the book emphasize geographical perspectives in planning and implementing public health interventions, whereas latter chapters focus more on program evaluation. Early chapters speak in general terms about the importance of space and place, and related concepts such as mobility and vulnerability in planning public health interventions for specific health issues. The central argument is that interventions should reflect the place-based processes that give rise to health inequalities and thus reflect distinctive population- and place-specific circumstances. In addition, the research conducted in these chapters illustrates the use of geographic data and methods in spatially targeting interventions on areas of high need and in supporting partnerships to develop effective intervention strategies. Chapters in the latter half of the volume build on these themes in addressing the *evaluation* of particular public health interventions, including bednets for malaria control, smoking cessation policies, and school nutrition programs. Through detailed case studies and natural experiments comparing health outcomes before and after intervention, the authors identify important place-based variation and sensitivity to local context that requires attention in the planning and implementation processes.

The organization of chapters also reflects our effort to highlight the richness and diversity of geographical contributions to PHIR. By juxtaposing chapters addressing specific health concerns – HIV/AIDS, vector-borne disease, obesity, and physical activity – we expose geographers' diverse methodological and conceptual perspectives on the causes of specific health concerns and the planning and implementation of appropriate and effective interventions. Varying scales of analysis, from global and regional viewpoints to the local spaces of daily life, reveal the range of social and environmental forces that shape people's health and constrain and facilitate intervention.

The first substantive chapter by Barb Riley, Kerry Robinson, Martin Taylor and Cameron Willis underscores the importance of partnerships in population health. This is an outstanding writing team in the area of PHIR, with two of the authors (Riley and Willis) affiliated with an academic think tank related to population health impact, a third (Robinson) in a senior policy position at the national public health agency in Canada and the fourth (Taylor) a career academic researcher and administrator who can be considered a pioneer in the subdiscipline of health geography. In this chapter, these authors draw on their tremendous experience as researchers to explore the relative contribution(s) of geography to three major research projects involving large-scale population health partnership initiatives across a range of spatial scales. These include: (1) the Community Intervention Trial for Smoking Cessation (COMMIT), a multi-site community-based trial; (2) the Canadian Heart Health Initiative, a 20-year program that involved partnerships between public health leaders and researchers; and (3) the Public Health Agency of Canada's (PHAC) multi-sector partnership program designed to promote chronic disease prevention and healthy living through novel public-private partnerships. This chapter provides useful insights on the role of geography and geographers in enhancing population health through partnerships.

Community partnerships, and their roles in shaping effective public health interventions, are a central theme in Olson *et al.*'s chapter on the Milwaukee Vacant Land Inventory. As in many cities in North America, Milwaukee's low-income neighborhoods are dotted with vacant properties – a legacy of deindustrialization and population decline. Olson and co-authors argue that these properties have complex implications for population health. On the one hand, they diminish opportunities for social interactions and engagement in communities that already face a myriad of health concerns, while at the same time, they represent enormous untapped potential for developing health-promoting spaces for recreation, relaxation, cultivation, and many other purposes. The authors discuss the development of a GIS-based inventory of vacant lots that incorporates both qualitative and quantitative indicators of attributes, quality, and land use for each property. A framework is presented describing opportunities for community involvement in directing the redevelopment process to support and foster community well-being. The use of GIS in cataloging and characterizing vacant properties and supporting community participation in developing interventions is an innovative feature of this chapter.

Geospatial data and methods can be important tools in developing and targeting public health interventions, especially when the data reveal patterns of ill-health that may be hidden from view. This key point is illustrated in Oppong *et al.*'s chapter which examines spatial and social inequalities in HIV/AIDS survival in Dallas County, Texas. Using data on survival time among people diagnosed with HIV/AIDS, the authors assess disparities in survival by race and ethnicity across groups of neighborhoods with different levels of socio-economic disadvantage. In addition to revealing racial disparities in survival, the data and analysis also dispute the concept of a strict socio-economic gradient in survival by indicating that those infected with HIV living in neighborhoods of moderate socio-economic status (SES) have shorter life spans than those living in higher or lower SES neighborhoods. The authors suggest that federal health programs such as Medicaid and Ryan White HIV intervention activities benefit those living in low SES areas and improve their health outcomes. Although the chapter does not investigate the interventions themselves, it highlights the importance of understanding linkages between social and spatial vulnerability in HIV/AIDS research and the need for data-driven approaches that may reveal people and places where interventions are most needed.

Following Oppong and colleagues' chapter, Nathaniel Lewis takes a critical lens to sexual health and HIV prevention for gay men and men who have sex with men (MSM) in newcomer communities. Lewis has undertaken a systematic review of interventions in North America, to assess the types of interventions being used in this population, their (perceived) effectiveness, and the degree to which interventions address elements of resettlement and the migration experience. First, by employing systematic review Lewis highlights the usefulness of this methodology for contributing to PHIR (as previously indicated by Potvin, 2012). He argues that while North American health practitioners have increasingly moved towards tailored interventions based on age and ethno-racial backgrounds, few address the role of migration experience beyond the related issues of culture and acculturation. Lewis approaches his review by recognizing that the migration experience includes other factors, including stress, isolation, and the forging of new social networks that can independently influence the context of sexual risk. This chapter provides a critical appraisal of the strengths and weaknesses of each study, and concludes with some reflections about how geographers can contribute to the development, implementation, and evaluation of interventions targeting sexual minority groups within immigrant populations.

Dickin's chapter presents an innovative vulnerability framework for understanding the health consequences of global environmental change and developing appropriate interventions. Capturing both a system's exposure to environmental change and its capacity to respond to change, vulnerability is a relational concept that links social and environmental systems from the local to global scales. Dickin argues that such linkages are critically relevant for population health research and intervention, particularly as we consider the health consequences of global climate change. Examples of vulnerability assessment are presented for dengue, one of the most rapidly expanding vector-borne diseases, and a disease strongly linked to increasing human mobility and global environmental change. Case studies of dengue in Malaysia and Brazil illustrate how data on susceptible populations, temperature, vegetation, and other place-based factors can be combined and mapped to reveal uneven patterns of vulnerability which can guide intervention efforts. With its emphasis on dynamic interactions between social and ecological systems, the chapter's place-based vulnerability framework has broad relevance for many areas of population health intervention research.

Janko and Emch's chapter highlights the importance of place in public health intervention through a quantitative analysis of the uneven effectiveness of malaria control strategies in the Democratic Republic of Congo (DRC). They begin with the essential point that evaluations of population health interventions often focus on the average or global effect, rather than considering place-based differences in impacts that reflect local population and environmental conditions. This point is illustrated via an empirical analysis of regional differences in the effectiveness of treated and untreated bednets for malaria control. Using a Bayesian multilevel modeling approach, the authors show that the associations between malaria incidence and community-level bednet coverage vary greatly from one region of the DRC to another. In some places, higher bednet coverage reduces malaria transmission, even among people who do not use bednets themselves. However, these spillover benefits are not evident in certain rural regions, especially those faced with political instability and conflict. Despite their importance, these kinds of place-based differences in outcomes and effectiveness are rarely addressed in population health intervention research. Beyond emphasizing place-based effects, this chapter illustrates how advanced Bayesian statistical methods can be used in mapping and analyzing geographic variations in public health intervention outcomes and impacts.

The chapter by Fuller and Hobin examines how natural experiments can be leveraged to advance our understanding of the effectiveness of population and policy-level interventions for improving health. They argue, in a similar vein to others (e.g. Petticrew *et al.*, 2005) that natural experiments and quasi-experimental research designs may be able to bridge the gap in the quantity and quality of evidence for understanding the determinants of health inequalities, thereby advancing PHIR. Particularly, they highlight the usefulness of natural experiments for assessing the implementation and impacts of interventions implemented in various settings that have the potential to address chronic disease. Natural experiments have an underlying geography, as interventions are often allocated to particular communities or jurisdictional boundaries, albeit these are generally out of the control of the researcher.

Fuller and Hobin discuss the importance of embedding research components within these interventions to rigorously assess how they work and identify areas for improvement. Their chapter identifies when it is appropriate to undertake a natural experiment study design, based on four fundamental criteria, and they provide recommendations for study designs. They have ambitiously included three empirical examples of natural experiment research from their own research programs, focused on the impacts of bike sharing programs, school-based physical education policy, and nutrition labeling in Canada. They conclude their chapter by reflecting on the challenges faced when undertaking this type of research, and synthesize some of their key learnings, including picking up on the discussion by Riley *et al.* in Chapter 2, highlighting the importance of stakeholder partnerships.

Scott Leatherdale is an Applied Public Health Chair in the School of Public Health at the University of Waterloo in Canada. The COMPASS study described in his chapter is one of the most innovative PHIR initiatives in the world today. As the title describes, COMPASS is a research platform for evaluating natural experiments related to youth health and health behaviors in Canadian schools. In so doing, this scientific endeavor intends to generate practice-based evidence (as opposed to evidence-based practice) through an innovative process of knowledge sharing and brokering. To date, a key barrier to population health prevention activities among youth in Canada has been the lack of systematically collected longitudinal data to inform and evaluate prevention activities in a comprehensive manner. This issue was raised almost a decade ago by Roy Cameron and his colleagues specific to youth tobacco control. To address this challenge, Leatherdale has been a pioneer in rethinking how applied primary prevention research among youth populations is both *conceptualized* and *conducted*. His chapter describes the conceptualization of COMPASS, the complicated data collection process, the innovative knowledge dissemination and brokering process, and the impacts on youth health in Canada. In so doing, he underscores both the importance of *context* or the (school and community) environments within which these behaviors happen, as well as the importance of partnerships between researchers and end users.

The chapter by Vine describes the undertaking of a process evaluation that explores the multi-level political contexts within which school nutrition policies are implemented. Specifically, this chapter evaluates the processes by which school nutrition policies attempt to change healthy eating behaviors in secondary schools, and tries to understand the contexts within which such policies operate. In doing so, Vine highlights the contribution of process evaluations to PHIR by seizing the opportunity to evaluate the implementation of a provincial school food and beverage policy in Ontario, Canada. Vine's findings highlight an area that is often overlooked in PHIR – the unintended consequences of (well-meaning) interventions. In particular, she indicates that the underlying economics of how food and beverages are provided in school settings (i.e. through cafeterias that are largely driven by revenue generation) may be problematic. She draws an important link between the school environment and neighbourhood context, and offers suggestions for using community-based partnerships as levers for developing equitable, sustainable, and effective interventions for healthy eating in schools in the future.

Leia Minaker and colleagues attack a major public health concern in Canada and elsewhere – intervening in the food retail environment. How do we provide access (geographical, financial) to healthy foods to a range of populations, including those most vulnerable? As these authors remind us:

> Food environments have been identified as a key predictor of people's ability to eat well and have consequences for overall population health and wellbeing. (Minaker *et al.*, this volume)

As such, their chapter focuses on healthy food retail interventions, particularly for populations at risk of developing nutrition-related chronic disease. This very useful addition to the volume first reviews a range of conceptual models used to frame (healthy) food environments, a comprehensive discussion of related concepts like food swamps and food deserts, and a review of population health interventions in the area of (healthy) food retail, including healthy corner stores.

Louise Potvin is a Canada Research Chair in Community Approaches and Health Inequality, who has written extensively on PHIR. In the concluding chapter of this volume, Potvin ties together some salient themes emerging from across the other chapters in the book, and explains the potential value added to PHIR by geographers. In this chapter, Potvin introduces some challenging ideas for population health intervention researchers in general, and geographers in particular.

Conclusion

> There has never been a more stimulating or crucial time to act. There is no point in changing thinking in population health, if we cannot change history with it. (Hawe and Potvin, 2009: I-13)

As research on population health increasingly transitions from documenting and understanding population health issues towards addressing health and health inequalities through intervention, we see this volume as a step towards defining an explicit place for geographers in PHIR. It seems obvious that this will relate to providing a rich understanding of the contexts, broadly defined, within which interventions are designed, implemented, monitored, and maintained. In the quote above, Hawe and Potvin highlight the critical juncture for PHIR, particularly given the early stages of its development as an emerging field of inquiry. We agree, and given the relative nascency of PHIR, we urge geographers to continue to claim a place in PHIR, and inject the same critical interrogation of the impact of place and space on interventions as they have previously done with population health.

2 Partnerships for population health improvement

Geographic perspectives

Barbara L. Riley, Kerry L. Robinson,
S. Martin Taylor, and Cameron D. Willis

Most health behaviors are influenced by a complex array of factors operating at multiple levels within a social ecological system, including individual, interpersonal, family, organizational, inter-organizational/network, community, and societal levels. Positive and widespread change in behaviors requires effective implementation of intentionally coordinated, evidence-informed interventions that promote behavioral and environmental changes at a population level (Green *et al.*, 1996; Edwards *et al.*, 2004). These population interventions span health and non-health sectors and multiple settings (e.g. workplaces, schools, communities) and jurisdictions (e.g. local, provincial/territorial, national). They also require coordinated efforts from research, policy, and practice perspectives. The design and implementation of population health interventions therefore require active engagement of organizations and sectors working together in different forms (Crilly *et al.*, 2010) and provide a rationale for the importance of partnerships (Best, Stokols, *et al.*, 2003).

The knowledge base on partnerships is eclectic and disparate. Specific to population health, calls for scholarship to understand and improve partnerships is growing (Hunter *et al.*, 2010). In particular, many questions remain about how the origins, form, function, and performance of partnerships are shaped by contexts, environments, politics, and demographics. Studying these questions from multiple disciplinary perspectives will provide the richest insights. The goal of this chapter is to explore how *geographic perspectives* may add value to programs of research and practice on partnerships for population health interventions.

Our approach was informed by a search of studies of health partnership initiatives in the geographic literature. The yield was scant; in the past five years less than ten peer-reviewed publications appeared in each of *Social Science and Medicine, Health and Place*, and *Progress in Human Geography*. All publications had a focus on partnerships in health but were not specific to population health interventions. The range of topics and jurisdictions was diverse, including studies of network theories of scale, healthcare delivery models, and engagement for local policy change, from high-, low- and middle-income countries. The results show that, while relevant to geographers, partnerships in health have not been a major area of recent inquiry.

The chapter draws on the experiences of the authors as researchers in three major research projects involving large-scale population health initiatives. The three projects represent significant experiments (a mix of natural and scientific) and investments at national and international scales, provide diverse examples of partnerships for population health improvement, and indicate the potential for contributions from geographic perspectives. The projects are (1) the Community Intervention Trial for Smoking Cessation (COMMIT), a multi-site community-based trial that included partnerships between the community and the research team; (2) the Canadian Heart Health Initiative, a 20-year initiative that involved partnerships between public health leaders and researchers; and (3) the Public Health Agency of Canada's (PHAC) multi-sector partnership program designed to promote chronic disease prevention and healthy living through novel public-private partnerships. Informed by these examples, the chapter concludes with a summary of insights on how geographic perspectives may add value to programs of research and practice on partnerships for population health interventions.

Partnerships for population health interventions

In this chapter, partnerships refer to structured forms of organizing for linking (1) research, policy and practice perspectives, and (2) two or more organizations from two or more sectors (e.g. public, private). Structured organizational forms typically refer to groups that are formally established, have a common goal, are governed and goal-directed, and are semi-autonomous from participating organizations (Provan *et al.* 2007).

Partnerships vary on many dimensions. For example, partnerships for population health interventions vary according to level of integration, on a continuum from low to high integration (Bailey and Koney, 2000; Gajda, 2004). Integration is low if the process and structure are limited to communicating information and exploring interests, and is often referred to as *networking* or *cooperation*. Integration is considered medium if the group plans together to achieve mutual goals while maintaining separate identities and is often described as *coordination*. Integration is highest if the group merges to form a single identity which is typically referred to as *collaboration*.

Partnerships for population health interventions also vary according to participation by different organizations and sectors. With respect to organizations, many types may be involved, such as universities, public health organizations, workplaces, health charities, municipal governments, among others. With respect to sectors, public, private, and charitable sectors are normally considered relevant to population health interventions. For example, Hunter *et al.* (2010) describe two broad types of partnership: public-public or public-private (with private meaning either for-profit commercial companies or not-for-profit third-sector voluntary bodies, or a mix of the two). Multi-sector partnerships have also been conceptualized as events within systems, and focus on the organizational relationships as well as the relationship between partnerships and their context (Woulfe *et al.*, 2010).

The prevailing orthodoxy has been that partnerships are intrinsically "a good thing" and a "must have" if only because many complex problems demand a cross-cutting approach if they are to be successfully tackled (Hunter *et al.*, 2010). But there has been little research examining partnerships concerned with public health, and they are arguably some of the most complex to understand and improve (Woulfe *et al.*, 2010), especially given the wide range of players typically involved, the multi-level and intersectoral nature of effective interventions, and an evidence base that may be partial, contested or even absent altogether.

A major gap in the literature on partnerships is a significant, and almost exclusive, focus on process issues rather than on outcomes. A recent systematic review concluded there is an absence of clear evidence of the effects of public health partnerships on health outcomes (Smith *et al.*, 2009). Echoing Dowling *et al.*'s (2004) call, the need is for research which seeks to explore the success of partnerships in effecting changes in service delivery and, if possible, to establish the subsequent effects on the health and well-being of a population. Efforts to answer this call have led to more questions about meaningful partnership outcomes or ways to judge the performance of partnerships. For example, Boydell *et al.* (2008) consider that partnerships should be valued from the perspective of "intangible assets." This led to the development of a model describing the benefits of partnerships in terms of how they improve health (Boydell and Rugkasa 2007). These benefits included: the connections made by partnership members, the learning that takes place, and the enhanced capacity to act as a result. In the context of this chapter, the question is how geographic perspectives may contribute to addressing this significant knowledge gap.

Geographic perspectives

Geographic perspectives on health and healthcare are quite eclectic in terms of the theoretical and methodological traditions on which they draw, marrying diverse social, economic, and ecological theories with a combination of quantitative (epidemiologic) and qualitative methods, reflecting the complexity of explaining and understanding place/regional variations in population health outcomes and healthcare systems. As has often been argued, it is perhaps its holistic and integrative perspective that best characterizes and distinguishes the geographic approach.

The concept of health and place is now a primary lens for conceiving and conducting population health research from a geographic perspective; population health outcomes are understood to be contextually related at a range of geographical (local and regional) scales; at the urban scale, for example, city neighbourhoods may serve as units of observation and analysis to investigate how the complex of cultural, social, economic and environmental factors function as co-determinants of population health; this approach is consistent with the eco-social theory of population health emergent in social epidemiology (Krieger, 2011).

While geographical models for population health interventions and the role of partnerships in particular have not been given the same attention, a *health and place* perspective argues for integrated system planning and delivery through multiple social agencies within various settings (schools, workplaces, healthcare settings, social services, etc.) and across various scales (e.g. local, provincial, national), working in partnership and engaging the publics they serve. The presence and potential of geographical perspectives are explored in the three cases that follow.

The Community Intervention Trial for Smoking Cessation

The Community Intervention Trial for Smoking Cessation (COMMIT) is an example of a multi-site community-based trial that included partnerships between the research team and the community. It is an example that in some ways "discovered" the importance of partnerships for population health interventions rather than partnerships being an explicit focus of inquiry. One of the authors (Taylor) was a member of the COMMIT research team and a geographer.

COMMIT was conceived and designed as a phase 3 (test of the effectiveness of an intervention in general populations) study by the US National Cancer Institute building on prior demonstrations of the efficacy and effectiveness of various community-based interventions to increase cessation rates among light to moderate and heavy smokers (COMMIT, 1995a). The principal purpose of COMMIT as a community trial was to determine the effect of a bundle of intervention strategies delivered concurrently in a set of intervention communities matched geographically and socially with a set of paired control communities. Eleven pairs were included, judged to provide sufficient power to detect intervention effects with the primary focus on quit rate changes among heavy smoker cohorts (>25 cigarettes per day) but with attention also paid to effects of cessation rates among light to moderate smokers (<25 cigarettes per day). In addition, prevalence surveys of smoking rates were conducted to determine changes in overall rates of smoking in the intervention vs control communities over the four-year duration of the trial which began in 1989 and ended in 1993 (1995b). A secondary objective of COMMIT was to examine changes in smoking norms and attitudes recognizing that attitude change could be a precursor of smoking behavior change (Taylor et al., 1998a; Taylor et al., 1998b). This was arguably all the more relevant because of the relatively short duration of the intervention (four years) and thereby the possible lagged effect of behavior change (relative to attitude change). The intention was to use the findings of the study to develop and implement a national implementation program as a phase 4 NCI initiative, subsequently named ASSIST. COMMIT was at the time one of the largest community trials ever conducted.

The results were modest. There were no attributable changes in quit rates among heavy smokers, but there was a significant reduction in cessation among light to moderate smokers that was judged to be of sufficient magnitude to have public health consequence and benefit (COMMIT, 1995a). Smoking prevalence rates declined over the period of the trial but the attributable difference between the intervention and comparison communities was low (COMMIT, 1995b).

Critiques of the study findings questioned the logic and value of the approach adopted in COMMIT, particularly its effectiveness to impact hard-to-reach and hard-to-influence heavy smokers (Susser, 1995). Moreover, the adoption of a design involving concurrent delivery of a suite of community interventions inevitably precluded parsing the results to determine relative attribution to particular intervention channels that included workplaces, physicians' offices, churches, schools, media, etc. In short, the strength of the COMMIT design – the bundling of previously proven effective interventions – and their concurrent delivery on a community-wide basis was also its weakness in preventing any clear determination of relative contribution to modest effects which might then inform the design of subsequent public health initiatives, including ASSIST as the national dissemination program following COMMIT (Fisher, 1995).

What is of particular interest and focus for this chapter is the inclusion of geographic perspectives in the COMMIT study. One of the authors (Taylor) was a geographer and COMMIT co-investigator as a member of the University of Waterloo-McMaster University team funded as one of the eleven for the study overall, and the only one in Canada, the rest being in the United States As part of the response to the COMMIT RFP, the Canadian team had to identify a pair of matched communities for inclusion in the study. This involved matching potential communities in southern Ontario based on population size, socio-demographic, and socio-cultural factors, largely derived from census data, and on smoking rates and policy. This was an intrinsically geographical exercise using cluster analysis to identify a best-matched pair which was determined to be Brantford and Peterborough. As part of the randomization for the trial design, Brantford was selected as the intervention community and Peterborough as the control. What is important here though is that for the main trial design and power calculations, the characteristics of individual pairs were factored out; the power of the trial depended on effect size calculated across all eleven pairs, such that differences within pairs or between pairs was essentially ignored. In short, the geography was discounted!

What is revealing is that *ex post facto* possible geographic effects and differences were given much more attention, albeit recognizing the statistical limits of the design to reach strong conclusions. That said, for each pair, the cohort sizes of heavy and light to moderate smokers (400 for each cohort in both communities in the pair) and the sample sizes for the pre (N = 5,400 per community) and post (N = 2,300 per community) prevalence surveys was by most standards substantial. Not surprisingly perhaps, primary interest in pursuing these follow-up questions was the purview of the social scientists among the COMMIT investigators, and especially the geographers in the Canadian team (Ross and Taylor, 1998).

The Canadian COMMIT team assumed particular responsibility for the study of attitude change (Taylor *et al.*, 1998a; Taylor *et al.*, 1998b). Two attitude scales were developed to tap two constructs: *smoking as a public health problem* and *norms and values concerning smoking*. While all 22 communities showed attitude change favouring non-smoking over the intervention period (1989–93), the incremental effect of COMMIT was modest and limited primarily to heavy smokers' beliefs about smoking as a public health problem, but even then the trial-wide effect across the eleven pairs of communities was not significant. That said, the COMMIT data provided a unique opportunity to examine attitudes towards smoking of a geographically diverse and dispersed (continent-wide, California to Massachusetts, plus Ontario) group of communities in 1989 before the start of COMMIT (Ross and Taylor, 1998). The potential insights to be gained were substantially enhanced by the concurrent completion of an extensive survey of state policies and tobacco control activities in that same year (Choi *et al.*, 1991). The analysis proceeded in two stages: an analysis of variance of regional and community level differences in smoking attitudes in 1989; and a subsequent qualitative study of selected communities showing large variation in attitudes, relating those differences to the past and current policy environment.

In brief, significant regional and community level differences were found for both attitude measures – *smoking as a public health problem* and *norms and values concerning smoking*. Based on between and within region variation as well as geographical diversity, five community pairs were identified for the second stage analysis – North Carolina, Iowa, Washington State, New Jersey and New Mexico – with the aim of exploring regional and local factors that might account for variations in smoking attitudes. The analysis identified several plausible explanatory factors including economic reliance on the tobacco industry, libertarian political orientation, socio-economic conditions, legislative context and ethnic composition (Ross and Taylor, 1998). The results supported public health efforts to reduce smoking rates that are targeted beyond individual smokers to encompass the broader social and policy environment.

Brantford, as one of the eleven COMMIT intervention communities, provided an opportunity to conduct a nested qualitative case study (Poland, 1993) focused especially on the socio-cultural environments, the social geography of heavy smokers supporting the perpetuation of smoking behaviour and thereby resistance to the types of community-wide intervention efforts implemented in COMMIT. This in-depth investigation pointed to the importance of social networks among heavy smokers, their largely self-created cultural support systems for continued smoking, and their ignorance of and/or resistance to COMMIT channels and activities in Brantford. The implication was to question the impact achievable by a community-wide intervention and the parallel need for more targeted strategies explicitly designed to reach heavy smokers, those most resistant to change. As such, the finding echoes an editorial commentary on the main outcome findings of COMMIT (Fisher, 1995): ". . . smoking has become a problem not of whole communities but of subgroups within those communities. It may be unrealistic for a standardized community program to reach important subgroups with high prevalence of health problems." Arguably this is even more true two decades later with the continued reduction in (heavy) smoking rates.

While now a somewhat dated example, the COMMIT study is instructive in several respects for the primary focus of this chapter on the importance of inter-organizational and multi-sectoral partnerships for achieving positive population health impacts.

COMMIT remains one of the largest community intervention trials thus far conducted focused on what is still among the greatest population health challenges, namely smoking behavior. More directly, it involved a complex set of concurrent interventions coordinated and managed at the community level. Implicitly, therefore, it required the coordination of activities delivered through diverse channels and partner organizations spread across several sectors. The fact that the study was designed and conducted as a trial was both its strength for experimental integrity and its weakness to meet the challenge of ensuring that the intervention protocol was sufficiently standardized as to not compromise the study design, while being appropriately customized to suit the realities of local communities as organic social systems. The implications for interpreting and understanding the study outcomes were anticipated in a paper focused on process evaluation in COMMIT (Corbett *et al.*, 1990). The following extract is particularly instructive:

> Some communities had a long history of coalition-building and grassroots activities, while others had little experience in working together as a community on a social issue. This 'experience factor' may have an influence on the length of time required for communities to mobilize around the smoking issue and may thus affect the trial outcome. Similarly, there appeared to be great diversity among the communities in the structures available to support and encourage smoking cessation. Some communities, for instance, had few smoking cessation providers and low visibility of the major health voluntary organizations that promote smoking cessation, while others had many more smoking cessation resources and service providers. Communities with few prior structures for provision of cessation services may have difficulty meeting the trial's early targets, if not the long-term cessation goals . . . The interpretation and generalization of results depends on reliable, valid monitoring of these influences and processes.

Leaving aside that this commentary reflects a past reality rather than current context for the communities in question, the basic point is just as salient: communities are complex social systems and population health interventions are perturbations whose ripple effects are difficult to predict, analyze, and therefore generalize. There are external and internal factors at play that cannot simply be written off as "noise" in the system that are sufficiently randomized so as not to distort the main intervention effects. Indeed, these background influences may be sufficient to dilute, diminish, or disguise the main effects. It follows that the importance is underlined of giving due consideration to regional and local variability as was the case for the study of attitude variations in COMMIT.

While not claiming primacy on the theoretical insights needed to design, implement, and evaluate population health interventions as large-scale and complex as those attempted in COMMIT, suffice it to conclude that geography matters. The geographic lens and thereby the perspective of (health) geographers are particularly valuable for studying eco-social processes in their regional and local contexts. The fact that geographers *per se* have been a conspicuous minority among investigators in major population health intervention studies, and to that extent have left it to researchers from other disciplines to "discover" and apply geographic approaches, is hardly the issue. The fact remains that a hallmark of geographic inquiry is the integrative analysis of process outcomes at the spatial scale of communities and regions, an intellectual competence and research skill which, as COMMIT well demonstrated, has particular force for population health intervention studies.

The Canadian Heart Health Initiative

The Canadian Heart Health Initiative (CHHI) represents a long-term partnership-based initiative across government and NGO sectors that integrated both intervention and geographical research perspectives on population-level approaches to prevent chronic disease. Three authors were directly involved as research team members of the CHHI demonstration phase (Riley) and the dissemination phase (Taylor, Riley, Robinson).

In Canada, the fight against chronic disease has been driven by national and provincial coalitions and alliances of civil, government, and professional organizations informed by experiences of large-scale community disease prevention trials and disease-specific prevention initiatives founded on inter-organizational partnerships (e.g. the Stanford Five-City Project and the North Karelia Project). The CHHI was a five-phase, 19-year undertaking during the period 1986 to 2005, aimed at the cardiovascular disease epidemic in Canada. The long-term goals of the CHHI were to reduce premature morbidity and mortality from heart disease and to reduce the prevalence of modifiable risk factors (e.g. smoking, physical inactivity, high blood pressure, unhealthy eating) (Health and Wellness Canada, 1987). In the short term, the CHHI focused on building capacity in the public health system (both nationally and in all ten participating provinces), delivering effective heart health interventions at the community level, and developing partnerships that would sustain efforts and impact. The overall initiative operated as parallel provincial intervention research projects that undertook strategies to accommodate differences in the respective provinces' size, population, health system structures, and economic and political contexts.

Although the CHHI began as a federal, provincial government and non-governmental organization partnership to enhance *heart health*, its research and intervention activities evolved over time and contributed to collaborative and integrated chronic disease prevention and healthy living promotion efforts across Canada (Robinson *et al.*, 2007a). The guiding principles of the CHHI included: (1) health program collaboration at national, provincial, and local levels; (2) recognition of the need to build health research and intervention capacity; (3) integration of programs into existing public health systems; (4) incorporation of population-based and high-risk approaches to programming; and (5) targeting of common chronic disease risk factors and interventions (Health Canada, 1992).

From 1986 to 2003, CHHI evolved through stages of national policy development, national risk factor surveys, provincial-level demonstrations of programs and interventions, and dissemination of programs and models developed through the demonstration phase (Figure 2.1). The theoretical underpinnings of the CHHI as a whole are based on the socio-ecological model of health promotion focused on improving the health of populations by influencing various social and physical environments to affect change at various scales from individual, to family, to community and society (McLeroy *et al.*, 1988).

Both the demonstration and dissemination phases of the CHHI integrated interventions and research in order to understand and test the implementation and impact of community-based health promotion programs and the development and transfer of capacity to implement complex population health interventions (e.g. workplace health programs, community trail changes). While each provincial project had its own research questions tailored to the nature of interventions and jurisdiction, synthesis and evaluation research undertaken across the projects used a geographical lens in order to examine: (1) the implementation of capacity building and dissemination interventions; (2) the levels of capacity and dissemination at multiple spatial scales (local, provincial, national levels); (3) contextual (organizational, local, and system) factors affecting dissemination interventions and research processes; and (4) facilitators and barriers associated with health promotion dissemination and capacity building (Elliott *et al.*, 2003). The synthesis and evaluation research across seven provincial dissemination projects used a two-phase, mixed-methods parallel case study and a cross-national survey to measure provincial variation in chronic disease prevention capacity and implementation by public health organizations (Hanusaik *et al.*, 2009; Masuda *et al.*, 2012).

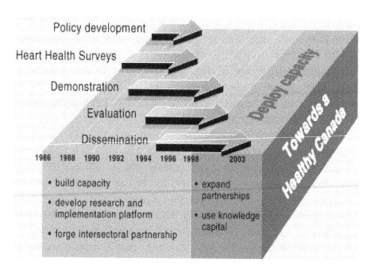

Figure 2.1 Phases of the Canadian Heart Health Intitiative, 1986–2005 (Public Health Agency of Canada Multi-sector Partnership Program).

Aside from the initial policy phase which was federally led, all other phases of the CHHI were jointly funded through federal and provincial government agreements. In order to receive federal funds, each province had to match federal contributions and resources were also leveraged from many non-government organizations. For example, in the demonstration phase, an $8.8 million dollar federal government investment leveraged over $24 million from provincial governments and other partner sources (Riley and Feltracco, 2002).

The CHHI represented significant collaboration between research, practice, and policy players across Canada. The model was founded on creating intersectoral and multi-disciplinary teams to build and apply the evidence base for community heart health promotion tailored to each provincial context. The CHHI was essentially a network of ten provincial projects. The intervention research questions of each project were driven by the evidence needs of policy-maker co-investigators and practitioners in each province. Some project teams were led by the provincial government (e.g. Ontario, Newfoundland and Labrador, Prince Edward Island), while others were led out of universities (e.g. Alberta, Manitoba, Quebec). Non-governmental organizations such as the Heart and Stroke Foundation of Canada were centrally involved at the outset of the CHHI at both national and local levels.

The nature and scope of partnerships during the CHHI evolved throughout its various phases and differed across diverse provincial health system contexts, thus their influence and impact on both organizational capacity and population health interventions varied over time and geography. Common areas of impact across the provincial projects and the national public health system related to: organizational capacity (knowledge, skills, resources), implementation of population health interventions, and integrated chronic disease prevention policy and strategy development.

During the Demonstration phase, coalition building engaged hundreds of multi-sectoral partners at local levels across the country from recreation, education, the private sector, professional associations, community-based groups, and volunteers to implement and test over 300 community-based prevention and promotion interventions in over 150 communities (Conference of Principal Investigators of Heart Health, 2001).

In the Dissemination phase, a pattern of modest, positive changes in organizational capacity for chronic disease prevention was documented with six of seven provinces studied documenting change in at least five of the seven capacity dimensions measured (Robinson *et al.*, 2007b). Further, cross-provincial qualitative and quantitative comparisons demonstrated that widely divergent dissemination strategies can each have considerable influence on the effectiveness of health promotion uptake at the provincial level. Six of seven provincial dissemination initiatives studied demonstrated increases in the level or comprehensiveness of chronic disease prevention initiatives implemented (across multiple risk factors/ determinants of health and settings) with evidence of progress on sustainability through increased partnerships and emphasis on intersectorality (Masuda *et al.*, 2009). Overall, the findings of this interpretive and multiple provincial case study support the notion that investment in capacity building, even limited human and financial resources, can produce organizational interventions that appear to have a positive influence on the capacity for and implementation of population-level chronic disease prevention (Robinson *et al.*, 2007b).

Finally, interview and document analysis findings revealed that seven provincial projects' intervention research results and coalition partnerships actively led or supported the development of province-wide chronic disease prevention and wellness initiatives that attracted provincial funding and ongoing technical support for local capacity building (Robinson *et al.*, 2007a). These strategies and policies helped fund the regional implementation efforts and the provincial coordination of chronic disease prevention efforts.

Partnerships throughout the CHHI were both supported and challenged by a number of noteworthy factors. Across the initiative, limited provincial-level financial and human resources and different levels of commitment became a significant barrier that precluded the participation of two provinces in the final dissemination research phase, and provincial government budget reductions limited the intervention component for a third provincial project. Two CHHI studies that studied geographically diverse communities (rural, urban, culturally diverse, differing socio-economic status) found that having dedicated/skilled people, commitment to partnering, common goals, leadership for collaboration, and sharing resources/expertise were facilitators for collaboration across organizations on heart health promotion, whereas lack of time, a negative political/economic climate, hierarchical organizational structures, mandate differences and issues of mandate protection/turf were identified as significant barriers (Robinson and Elliott, 2000).

The Dissemination phase of the CHHI engaged several geographers as investigators in both the Ontario Dissemination Project and the subsequent pan-Canadian evaluation and synthesis study undertaken of seven provincial dissemination projects. While these two research projects explicitly integrated geographical perspectives into the CHHI, the very nature of the CHHI implicitly applied issues of geography as an initiative that focused on population-level community-based interventions aimed at changing social and physical environments for chronic disease prevention across three scales of the Canadian public health system – local, provincial, and national.

The scale (size, scope, level) of public health systems and dimensions of issues such as heart health or chronic disease prevention are socially constructed and, as in the case of health system structures, are regularly subject to revision (Frankish *et al.*, 2007). Nearly all provinces' health systems underwent a significant reform (devolution or centralization) during the CHHI, all of which had direct impacts on provincial project interventions, public health structures, and priorities. The influence of health system changes on interventions and research findings was a theme of study across many of the provincial projects. Provincial health policies corresponded to decisions about the appropriate level(s) at which healthcare should be managed in relation to efforts to achieve sustainability – either at the individual level or the collective/community level, or a combination of both

(Masuda *et al.*, 2012). Provincial CHHI projects developed health promotion and capacity-building strategies that variably supported or resisted the contexts they operated in, while allowing them to maintain resources and interventions to advance chronic disease prevention.

The importance and meaning of space and place underlies much of the partnership, intervention, and research of the CHHI. It was apparent through cross-case comparisons that differences in strategies adopted by provincial teams were a result of tailoring approaches based on the situational expertise of local health system contexts. For example, in the case of one resource-poor province, the research team took advantage of existing regional coalitions closely linked to regional health structures to embed their dissemination project into the existing infrastructure and encourage the uptake of program strategies by community volunteer groups. In contrast, the initial instability of another province's regional health system led their team to work outside the system and facilitate the creation of new community committees supported by facilitators. Thus a key message is that one size does not necessarily fit all in terms of best practice in health promotion dissemination (Masuda *et al.*, 2009).

The value of the CHHI as an initiative from which to draw insight for partnership approaches to population health interventions is that the diversity of contexts and intervention approaches used within and across provinces supports confidence in identified common patterns and findings. Moreover, the place-sensitive analysis of the development and evolution of partnerships, capacity-building models, and varied intervention impacts allows for richer understanding of approaches to support adaptation to other contexts. It is clear that the overall strength of the CHHI was the flexibility built into the initiative design, which allowed provincial partners to adapt their dissemination strategies in ways that were best suited to the heterogeneous and dynamic provincial environments in which chronic disease prevention was situated (Masuda *et al.*, 2009). The model of partnership building adopted by the CHHI based on co-investment, multi-jurisdictional involvement, respecting context, and engaging and building on existing community interest and capacity, may continue to hold tremendous value for chronic disease prevention and beyond.

The Public Health Agency of Canada (PHAC), Canada's federal public health body, is taking a leadership role in fostering, implementing, and learning from multi-sectoral partnerships for preventing chronic diseases and promoting healthy living (Public Health Agency of Canada, 2014). Three authors are involved in this initiative: Willis as the primary researcher, Riley as a research collaborator, and Robinson as a policy research leader within PHAC.

PHAC's multi-sectoral initiative is part of a broader interest of the government of Canada in the role of social innovation, highlighted in a recent Speech from the Throne that called for action on "opportunities presented by social finance and the successful National Call for Concepts for Social Finance." Through PHAC's multi-sectoral initiative, social innovation is being fostered through cross-sectoral engagement (Le Ber and Branzei, 2010) which seeks solutions to pressing social problems in ways that are more "effective, efficient, sustainable or just" than existing efforts (Phills *et al.*, 2008). Such an approach recognizes that no single organization or sector has the responsibility or capacity for addressing population and public health issues, and that it is only through inter-sectoral collaboration that lasting advancements in the health of people and places may be realized (Woulfe *et al.*, 2010; Provan and Lemaire, 2012). Through partnering with a range of key stakeholders, including other governments and departments, the private sector, charitable sectors, and not-for-profit organizations, PHAC is aiming to stimulate social innovation, support the implementation and evaluation of population health interventions, mobilize knowledge across sectoral boundaries, leverage new resources that expand the reach of programs to Canadian communities, and coordinate diverse efforts across levels of government.

PHAC's multi-sectoral approach to partnering is now providing support for multiple partnership projects, which vary in their geographic focus, scale, scope, purpose, partners and level of financial support. The goal of this program broadly is to promote healthy living and chronic disease prevention through the development of interventions that address common risk factors. Each project brings together organizational partners from across the private and for-profit sectors, academic and research institutions, non-government organizations, and private foundations, and can include Provincial and Territorial governments. A key component of PHAC's multi-sectoral initiative is the engagement of *non-traditional* partners, including private for-profit organizations and those working outside of the health sector, such as Loyalty One Inc. (Air Miles for Social Change), Lift Philanthropy, the Canadian Broadcasting Corporation (CBC) and the Canadian Football League. To navigate the development of these complex partnerships, PHAC has developed a range of tools, including a Partnership Guide (developed by the Centre for Chronic Disease Prevention), and partnership-specific measurement and evaluation tools. Guiding PHAC's innovation in partnership development is a commitment to protecting the public interest, fairness and impartiality, accountability, and vigilance to ethical risks.

The evolving platform of projects that form PHAC's multi-sectoral initiative provides a unique natural experiment for understanding inter-organizational partnerships for population health interventions in a diversity of contexts, communities, and conditions. The unique position of the government as a convenor of novel multi-sectoral partnerships is allowing PHAC to co-create a program of research – termed a Learning and Improvement Strategy – that will contribute new practical and scientific knowledge to the field of inter-organizational partnerships. This strategy is a deliberate effort to capture the unique learnings emerging from PHAC's experiences in brokering, implementing and sustaining multi-sectoral partnerships, and is built on the core principles of relevance, responsiveness, accuracy, and continuous improvement. In the context of PHAC's multi-sectoral partnerships, learning is thought to occur at different organizational levels (involved organizations, projects, and the initiative itself). With this diversity comes a need to promote widespread and continuous processes to enhance an organization's ability to accept, make sense of, and respond to internal and external change. This type of learning requires processes for systematically collecting and integrating information, collectively interpreting that information into new knowledge, and spurring collective action and experimentation to improve structures, processes, and outcomes (Business Dictionary, 2014).

The initial design for a learning and improvement strategy is based on a learning cycle incorporating six components, which captures information on a set of guiding questions. This cycle has been designed as a flexible, iterative and ongoing approach to learning that implements multiple, short-term cycles throughout the course of an 18-month to two-year period. The rationale behind these learning cycles (estimated to be 3–4 months in duration) is to be responsive to time-sensitive issues, allow ideas and solutions to emerge over time, engage relevant subsets of projects and teams, and make best use of existing resources and structures. Each cycle is based on a prioritized learning need, collectively identified and agreed upon by PHAC team members. Initially, these learning needs are focused on program reach, unintended and intended consequences of the multi-sectoral approach, and improving internal organizational design for enabling partnerships.

Following prioritization of learning needs, the early phases of each learning cycle aim to ground all activities in existing evidence, both conceptually and methodologically. This involves identifying key guiding frameworks and approaches (such as Collective Impact (Kania and Kramer, 2011), Systems Change (Foster-Fishman *et al.*, 2007), and the Grounded model for analyzing formation in cross-sector partnerships (Seitanidi *et al.*, 2010)) as well as specific methodological tools (e.g. the Collaboration Evaluation and Improvement Framework (CEIF) (Woodland and Hutton, 2012)) for gathering data. These tools are used to gather relevant information, often from a range of sources such as existing performance monitoring data and project reports, as well as from new sources such as interviews, focus groups, surveys, systems mapping and network analyses. Through rigorous analytical lenses, such as realist evaluation, these diverse sources of information are synthesized to generate a context-sensitive understanding of what works, for whom, in what settings, and why (Pawson, 2013). Evidence synthesized through this lens is then fed back to appropriate PHAC teams in ways that maximise engagement, including in-person learning workshops, interactive webinars, and tailored written products. Development of these products coincides with internal PHAC decision-making processes, allowing plans, templates, tools, and reporting to be informed by relevant evidence. This approach helps increase the level of interaction between PHAC staff and their own learning information, as well as optimize the use of that information in ongoing center planning and evaluation activities. Insights gained from the learning and improvement strategy therefore have value for the broad field of population health intervention research, as well as other government departments which are implementing or considering a multi-sectoral approach.

Given PHAC's current investment in a learning and improvement strategy, there is an opportunity to understand how a geographical lens may be applied and the value that may be gained from this perspective. For example, innovative insights may be gained into how multi-sectoral partnerships influence (and are influenced by) place, space, and scale. Given the diversity of examples offered by PHAC's multi-sectoral initiative, timing is ideal for incorporating these (and other) issues into ongoing data collections and analyses. For example, as part of the Play Exchange – a partnership involving a not-for-profit social enterprise (LIFT Philanthropy), a private-sector retail company (Canadian Tire), Canada's public broadcasting corporation (CBC television), and the Canadian federal government (PHAC) – individuals and communities across Canada are identifying innovative healthy living ideas from a diverse set of non-traditional sectors. Other projects, such as Lifestyle Prescriptions and Supports to Reduce the Risk of Diabetes in Rural and Remote Communities, more explicitly identify and target the health needs of populations from varying geographical locations. This project's focus on adapting and scaling up the Health-e-Steps model in rural Canadian communities provides solid foundations for applying a geographical lens to better understand how a multi-sectoral approach may be adapted to different contexts, populations, and infrastructures. Given the size and scope of these multi-sectoral projects, their effects (both positive and potentially negative) may be felt on groups and in regions far removed from their intended impacts. Understanding these impacts may be enhanced through a flexible and organic approach to how the borders and boundaries of a particular intervention are considered, and the populations, groups, and individuals who are ultimately affected. In line with PHAC's initial focus on reach and impact, understanding these often fuzzy boundaries is a key area for exploration in future work, and for which a geographical lens will add significant value.

Summary and conclusion

A core premise of this chapter is that partnerships are becoming increasingly important vehicles for designing, implementing, and sustaining population health interventions in Canada and abroad, yet our understanding of how to optimize their operations and impact remains relatively under-developed (Smith *et al.*, 2009; Woulfe *et al.*, 2010; Willis *et al.*, 2012; Willis *et al.*, 2013). The three case examples of COMMIT, CHHI and PHAC's multi-sector partnership program provide diverse examples of partnerships for population health initiatives and indicate the importance of incorporating geographic perspectives.

COMMIT, as the first example chronologically, used the most traditional design scientifically. COMMIT and other multi-site community-based intervention trials (e.g. cardiovascular disease prevention programs worldwide) essentially aimed to scale up study designs and methods from the clinic to the community. Partnerships between research teams and community members were mainly viewed as a way to assist with study implementation – to ensure evidence-based interventions were implemented with fidelity. A striking insight from COMMIT and its sister trials was the importance of community engagement, including a wide range of partnerships, not only for the purpose of implementing specific interventions, but also as an intervention approach within a socio-ecological system. In parallel, the significance of geographic perspectives emerged. COMMIT was an early example of surfacing the need for a place-based, context-sensitive understanding of health behaviors and of partnerships across organizations and sectors.

The CHHI took a fundamentally different approach compared to the previous generation of community-based trials. At its core, the CHHI was a capacity-building initiative. It was designed to build capacity across diverse settings and jurisdictions for the effective design, implementation, and evaluation of population-based interventions, and a partnership approach was central to accomplish these goals. Geographic perspectives became increasingly explicit in the later dissemination phase, and emphasized examining the influential role of many dimensions of context (e.g. physical, cultural, demographic, historical) on partnership development and intervention outcomes.

PHAC's multi-sector partnership program is poised to make significant contributions to understanding and improving partnerships, with a special focus on public-private partnerships. The emerging learning and improvement strategy will seek to answer context-sensitive questions about what types of partnerships work for whom and under what conditions. These questions are inherently place-based and represent an important opportunity for geographic inquiry.

Within the geographic discipline, while some examples exist of partnerships that have been viewed through a socio-geographical lens (Duff, 2011; Giles-Corti and Whitzman, 2012) there have been few efforts to explicitly apply this perspective to partnerships for population health interventions, leading to limited knowledge of how partnerships vary by place, space, and scale. Integrating this perspective into a broader understanding of partnerships is an important step toward a deeper and richer knowledge base on how, why, for whom, and where these partnerships have most impact, and the processes by which diverse actors and agencies can collaborate for population health improvement. Therefore this chapter is a call to geographers to consider programs of research on partnerships to strengthen population health, and to non-geographers who study partnership-based initiatives to incorporate geographic perspectives in their work.

3 The vacant land inventory

An approach to support vacant lot
redevelopment for population health
improvement in Milwaukee, WI, USA

*Jared Olson, Sandra Bogar, Kelly
Hoormann, Moriah Iverson, and
Kirsten Beyer*

The keen observation of Frank Zeidler, Milwaukee's long-standing mayor, that "Good leadership cannot come out of bad housing," is as true now as it was then (Gurda and Looze, 1999). The concept that where we live matters is now a central tenet in both public health and critical geographic scholarship. Researchers are examining and measuring how the structural, physical, and social features of urban environments impact population health (Frank and Engelke, 2001; Kuo and Sullivan, 2001; Baum and Palmer, 2002; Branas et al., 2011; Branas, 2013). At the same time, the spatial distribution of health inequities such as poverty, crime, unemployment, and chronic disease have deservedly received scrutiny (Morenoff et al., 2001; Brulle and Pellow, 2006; Marmot et al., 2008). Furthermore, research has suggested that the spatial patterns of disadvantage are the result of historical processes of racial segregation and economic disinvestment from inner cities (Margulis and Kenny, 2001; Acevedo-Garcia et al., 2003). Disparities in health and well-being observed between neighborhoods have focused attention on the spatial scale of the neighborhood, and researchers have identified neighborhood-level interventions as important assets for the improvement of community health (Sampson et al., 2002; Sampson et al., 2005; Moore and Diez Roux, 2006; Branas and Macdonald, 2014). Building on these suggestions, neighborhood interventions can be further enhanced and sustained when communities are actively engaged in planning and decision-making (Minkler, 2000).

Vacant lots: reflecting and reinforcing neighborhood health

Vacant lots, too often ignored, present a unique opportunity for neighborhood-level intervention. Indeed, vacant lot transformations designed to address urban public health problems have recently gained momentum (Branas et al., 2011; Garvin, Branas et al., 2013; Garvin, Cannuscio et al., 2013). In the depopulated centers of America's legacy cities, vacant lots, like boarded windows or litter, are commonly considered just another feature of neighborhood blight and disinvestment. Concentrated and clustered in neighborhoods already challenged by economic disinvestment, unemployment, and segregation, the ubiquity of vacant lots can mask their potential as a resource (Gorman *et al.*, 2001; Branas *et al.*, 2011). At their worst, vacant lots are associated with crime and violence, act as dumping grounds, and attract social disorder (Covington and Taylor, 1991; Liggett *et al.*, 2001; Branas *et al.*, 2011).

Vacant lots beset by social disorder reflect, reinforce, and reproduce disorder at the neighborhood level. As Figure 3.1 demonstrates, high-density concentrations of vacancy are a major challenge in Milwaukee's central city, also challenged by poverty. Often the history of a vacant lot can reveal something of the history of the neighborhood. A vacant lot may have previously contained an abandoned home, razed because it harbored undesirable activity. The loss of employment opportunities and the rise of poverty and crime may have led to the home's abandonment; perhaps it was a family home that was foreclosed upon. The exodus of one family from one home becomes the depopulation of an urban neighborhood at aggregate scales. The creeping poverty, outmigration, and social disorder associated with vacant spaces in disrepair feeds back into the neighborhood reputation, leading to further depopulation and divestment that only exacerbates disparities (Ross and Mirowsky, 1999; Sampson, 2009).

Given the strong link between health and socio-economic status (SES), it is not surprising to find that spatial patterning of health measures often reflect spatial patterns of SES (Diez-Roux, 1998; Schulz *et al.*, 2002). The spatial patterning of health suggests that both social determinants and neighborhood effects strongly influence individual health outcomes (Cohen *et al.*, 2003; Frieden, 2010). Vacant lots do not fit neatly within individual behavior models of public health; however, a social epidemiological perspective cognizant of these environmental influences provides a link between healthy neighborhoods and healthy individuals. Social epidemiology, like health geography, explores the relationship between place and health. Everything from national policies to city geographies shapes the contextual environment individuals inhabit. This context, be it familial, neighborhood, nation, or ecosystem, is embodied by the individual. The embodiment of the environment by the individual is the manner by which the deleterious effects of the neighborhood become the disease and illness of the individual. When these neighborhood effects are embodied by residents, spatial patterns of disparity and disadvantage emerge (Krieger, 2001a, 2001b, 2012a).

The potential of vacant lots

If health-damaging place features can be embodied, so too can health-promoting features. Community organizations and residents working towards neighborhood revitalization are beginning to paint a different picture of vacant lots. This picture concentrates on the untapped potential of what vacant lots *could* and *can* be, rather than what they have been. Vacant lot transformations have the potential to improve neighborhood reputations and promote additional community investment as they become community gardens, new homes, sources of employment, or imaginative multi-purpose combinations. Furthermore, if vacant lots are mobilized as assets, they can directly contribute to improved neighborhood health and

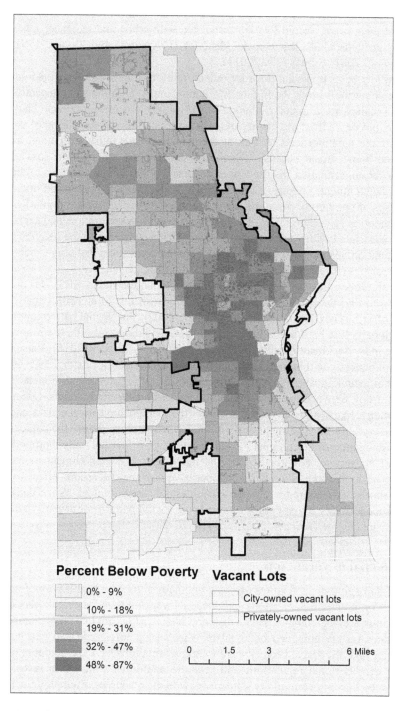

Figure 3.1 Vacant Lots, City of Milwaukee, WI, USA (May 2014)

also help mitigate persistent health disparities. This process is already underway in America's legacy cities as vacant lots are converted into community gardens, outdoor play and gathering spaces, and public art installations (Branas *et al.*, 2011; Park and Ciorici, 2013; Cleveland Neighborhood Progress, 2014; Detroit Future City, 2014; Pennsylvania Horticultural Society, 2014).

Even city stabilization and turf grass greening of vacant lots can have beneficial effects. Recent research demonstrates that converting a vacant lot into a green space can increase community residential property values and household wealth (Bucchianeri and Wachter, 2007; Heckert and Mennis, 2012; Bucchianeri *et al.*, 2012). Vacant lot greening has also been associated with reducing some types of crime such as gun assaults, as well as increasing physical activity while decreasing overall stress (Branas *et al.*, 2011). Additionally, green space, often in short supply in disadvantaged communities, has a number of wide-ranging positive health effects that could be promoted through vacant lot conversion (Kaplan, 1995; Kuo and Sullivan, 2001; Groenewegen *et al.*, 2006; Maas *et al.*, 2006; Alcock *et al.*, 2013; Beyer *et al.*, 2014).

From a public health perspective, the transformation of vacant lots remains nascent, but because vacant lots represent the intersection of structural, social, and environmental aspects of health, numerous disciplines inform the potential causal pathways and theoretical underpinnings behind vacant lot transformations. For example, the green space literature and the field of health geography provide key insights for vacant lot transformation by theorizing that place can support health by serving as a therapeutic landscape (Gesler, 1992; Kearns and Moon, 2002; Macintyre *et al.*, 2002; Cummins *et al.*, 2007). In addition, social capital, social cohesion, and collective efficacy each serve as indicators of health measures to document how physical transformations of the environment can improve community health (Jacobs, 1961; Kornhauser, 1978; Bursik, 1988; Sampson and Groves, 1989; Browning and Cagney, 2002; Groenewegen *et al.*, 2006; Eriksson, 2011). Healthy neighborhoods have assets such as healthy norms and values, a positive reputation, and healthy environments for home, work, and play that manifest positive health in their residents (Macintyre *et al.*, 2002).

From a structural perspective, the arrangements of power surrounding vacant lots are also changing and evolving. While vacant lots have traditionally been the purview of developers and city officials, the surfeit of lots has changed attitudes and opened space for greater community involvement with vacant lots (Burkholder, 2012; City of Milwaukee Department of City Development, 2013). This policy change marries well with increasing calls from public health researchers to include community engagement in public health interventions in order to increase intervention sustainability and build community capacity (Minkler, 2000). Asking how vacant lots can and should be transformed in the communities that live with vacant lots supports the community agency needed to transition deleterious spaces into salutogenic spaces aligned with local needs and preferences. A participatory geographical information system (GIS) that privileges local knowledge and values multiple perspectives is a key tool for community involvement and partnership, especially when situated within a grounded visualization framework that integrates GIS and ethnographic qualitative research (Cope and Elwood, 2009; Knigge and Cope, 2009). The recursive and iterative nature of grounded visualization, which uses qualitative data to inform quantitative data and vice versa, helps ensure that community voices are not only *heard*, but are also actively *integrated* into both the data collection and any transformations (Dunn, 2007).

Objective

The combination of potential and mutability make vacant lots a compelling focal point for health intervention. Knowing that neighborhoods have a profound effect on the health and well-being of community residents, vacant lots emerge as key leverage points for determining what those neighborhood effects will be. At the present time, few tools and processes have been developed to facilitate scientific research concerning vacant land transformations and the measurement of these transformations as public health interventions. Further, little work has focused specifically on engaging communities in designing the future of vacant land. Consequently, we sought to design a vacant lot inventory system that could be adapted and used by communities to capture the current characteristics of vacant lots and the spatial distribution of vacant lots. The vacant lot assessment tool was created to capture data in several important categories: economic characteristics of the lots, formal and informal uses of the lots, the environmental conditions of the lots and surrounding areas, and the lots' effect on the neighborhood through both material and cultural pathways. We theorized that these characteristics and contexts would provide valuable insights concerning how lot conversions can have the greatest impact on population health. This work should inform decision-making about lots and provide a model for future collaborative participatory GIS studies in conjunction with communities. Public participatory GIS (PPGIS) processes that gather volunteered information create a richer geospatial inventory that both informs health-focused vacant lot interventions and engages residents as partners in lot conversion design, implementation, and maintenance. This engagement makes vacant lot transformations more impactful and sustainable, and this successful cooperation to create neighborhood health assets can be subsequently shared as models through the same vacant lot inventory.

Methods

Study area

The study area for this work is defined as the Harambee neighborhood, located on the northeastern side of the city of Milwaukee, WI, USA. Harambee is one neighborhood within Milwaukee's central city, an area that has been historically Black. It has also seen the same challenges that other American legacy cities have faced: concentrated and persistent poverty, deindustrialization, segregation, and outmigration. These historical processes have created incredible disparities between the urban and suburban Milwaukee neighborhoods (Gurda and Looze, 1999; Chen *et al.*, 2011). To develop and pilot the inventory system, the Harambee neighborhood was selected for several reasons. First, the Harambee neighborhood is one of several city-defined neighborhoods in Milwaukee with a high density of vacant lots. Second, the neighborhood has relatively low SES and other structural challenges common to Milwaukee's central core. In addition, the study team has partners in the area and a priori knowledge of ongoing vacant lot redevelopment efforts in the neighborhood. An awareness of the existence of vacant lot redevelopment in the neighborhood was considered beneficial for the development of the system, as it can illustrate diversity in the use of vacant land and the functionality of the tools employed. Finally, these active redevelopment efforts suggest a dynamic and engaged citizenry with potential to contribute to public participatory vacant lot assessments in the future.

Defining vacancy

In Milwaukee, vacant lots are those individual lots, either publicly (city) or privately owned, that do not have any permanent enclosed structures. In practice, most of Harambee's vacant lots are lots that have previously had residential or commercial buildings that were razed as part of neighborhood stabilization efforts. Additionally, some vacant lots are residual parcels where the size, shape, or slope of the lot made construction impractical. The City of Milwaukee owns and assumes responsibility for the publicly owned vacant lots, most often acquired as distressed properties through tax foreclosures. Distressed properties owned by the city with structures that cannot be rehabilitated are stabilized as vacant lots (City of Milwaukee, 2013). Conversely, privately owned vacant lots, owned by individuals and not the municipality, are more heterogeneous in character and not subject to the same stabilization program the city implements. As a result, private vacant lots may not be viable targets for conversion and public health intervention when compared to public vacant lots.

Study sample

The city of Milwaukee maintains a substantial and accessible parcel database on its own website (City of Milwaukee, 2014a), and the Milwaukee Property Record (MPROP) database is also available from the City of Milwaukee website (City of Milwaukee, 2014b). These sources were the foundation for the study sample we examined in Harambee. After downloading the parcel and MPROP databases, all parcels within the Harambee neighborhood were extracted (n = 3,292). Vacant parcels (Land Use Code = 8880, n = 536) and vacant parcels with improvements (Land Use Code = 8885, n = 8) were identified. In order to determine whether vacant lots were city-owned or privately owned, the MPROP database was queried to identify all terms that indicate city ownership. These terms included: City of Milwaukee, City of Milw, City of Milw Redev Auth, Redevelopment Authority of City of Milw, and The Redevelopment Authority of the City of Milwaukee. Privately owned lots were flagged (private = 1, n = 130) and city lots were not (private = 0, n = 414). The full database was then joined to a database of properties maintained by the City of Milwaukee. Lots maintained by the city were flagged (CityMaint = 1, n = 399), while parcels not on the list were not flagged (CityMaint = 0, n = 145). The full sample of lots (n = 544) fell into four categories: (1) vacant lots with improvements (n = 8, 2 = city owned, 6 = privately owned); (2) vacant lots described by MPROP as city owned and included in the city maintenance database (n = 397); (3) vacant lots described by MPROP as privately owned and not included in the city mainte-nance database (n = 124); and (4) lots listed as city owned that were not included on the city maintenance list (n = 15). No lots were listed as privately owned in MPROP but maintained by the city. The study sample includes a 10 percent sample from categories 2 and 3. Lots with known improvement activity were excluded from this exercise, as they are fewer in number and may vary substantially from vacant lots with unknown activity. The full set of lots examined includes 40 city owned lots maintained by the city, and 12 privately owned lots, for a total sample size of 52 lots. We examined only a sample of lots to enable depth in characterizing vacant lots through direct observation.

Secondary data sources

Several existing databases were integrated to characterize the lots sampled. The Milwaukee Property Record Database (MPROP) is a parcel database for the City of Milwaukee. Originally constructed in 1975, this tabular database includes over 90 data variables that describe approximately 160,000 properties in the city of Milwaukee and can be easily linked to GIS boundaries of city parcels by tax key. The data is well-used by city departments and is updated daily (City of Milwaukee, 2014a). The EPA EnviroAtlas database is a collection of interactive tools and resources that facilitates the exploration of ecosystem services across the USA (Environmental Protection Agency, 2014). As part of this effort, EnviroAtlas staff have created one-meter landcover files for selected municipal areas throughout the country. Milwaukee, WI is one of the selected communities for which this data is available. Data files were obtained from EnviroAtlas staff (Jackson, 2014). The City of Milwaukee also maintains a database used by city officials who are tasked with maintaining the city's vacant land parcels. City officials shared this database with us for the purposes of this project (Brown, 2014). In total, the completeness and accessibility of the City of Milwaukee property records and EnviroAtlas data was a remarkable asset, one that may not be generalizable to other cities.

Systematic observation

Development of a vacant lot assessment tool

To enhance the information gathered from secondary databases with information about the use, quality, and features of lots, the project team developed a brief field assessment instrument. The goal of the instrument was to provide information that was not attainable through review of existing databases. Numerous audit tools currently exist for neighborhood assessments, but very few have been specifically targeted for the assessment of vacant lots, with the exception of one recent report (Kremer *et al.*, 2013). The project team reviewed existing neighborhood assessment instruments to identify questions and features of neighborhood environments that may be relevant in the context of vacant and redeveloped formerly vacant lots. Through this review, the team constructed an initial survey tool. Two team members then independently administered the tool on a sample of ten lots not included in the final sample. Assessors discussed their findings and the project team subsequently modified the tool to shorten and focus it on elements of primary interest and relevance.

The final tool included: (1) the Kremer assessment tool designed to determine actual use of the "vacant" space; (2) whether the lot appeared to be publicly or privately owned and managed; (3) the assets identified in the space (e.g. benches) and the quality of maintenance of those assets; (4) whether (and what type of) food was grown on the lot; (5) the quantity and type of litter observable on the lot; (6) the presence of graffiti and its message; (7) the proportion of nearby structures that appeared to be abandoned or uninhabitable; and (8) the assessor's opinion on whether the lot "strengthens" the neighborhood and their reasoning for this assessment. Finally, assessors were equipped with the ability to take photographs and video of the lot, the block, and any graffiti that was identified. The technology employed allowed for GPS tagging of assessor locations, and assessors entered tax keys for each lot assessed, to ensure that data was attributed to the correct lot.

Selection of a technology for field data collection

The project team reviewed a number of potential technologies to use in assessing lots, including Survey Monkey, GIS Cloud, Esri ArcGIS Collector, Harvest Your Data, and Fulcrum (ESRI, 2014; Finley, 2014; GIS Cloud, 2014; Harvest Your Data, 2014; Spatial Networks, 2014). Priorities for the tool included its ability to identify the geographic location of the participant, the ability to incorporate conditional questions to reduce redundant questioning, and the functionality and cost of the software. Fulcrum was selected for this project because of its user friendly interface, effective GPS tagging, ability to include cached map images and enter data offline, and reasonable price for a small number of users. No commercially available applications were found to meet all of the research team's priorities, and while Fulcrum was effective for a small number of assessors, its price scale depends on the number of users of the mobile device application, making it prohibitively expensive for participatory GIS endeavors. The challenges and opportunities of mobile GIS technology will be further addressed in the discussion.

Data and analysis

Table 3.1 provides a list of variables included in the database for each sampled vacant lot. These variables come from four key databases – the Milwaukee

Table 3.1 Variables included in database (parcel level)

Source	Variable	Description
Milwaukee Property Record (MPROP) Database	TAXKEY	Unique 10-digit ID assigned by the City
	CALAND	Current assessed land value
	CAIMPRV	Current assessed improvements value
	CATOTAL	Current total assessed value
	VALUEDLAND	Whether or not the lot has been assessed to have value (based on CATOTAL)
	CAEXMTYPE	Code indicating reason for property tax exemption (e.g. Playground/Tot Lot)
	CONVEYDATE	The year and month of the last real estate conveyance transaction
	OWNERNAME (1,2,3)	Name of the legal property owner
	NEIGHBORHOOD	City-defined neighborhoods
	LOTAREA	Size of the property in square feet
	ZONING	The current zoning of the property
	LANDUSE	Detailed land use codes (e.g. 8880-Vacant)
	LANDUSEGP	Category indicating general use of land (0-13)
	TAXDELQ	The number of years for which there are delinquent taxes due
	PRIVATE	Publicly or privately owned
EPA EnviroAtlas	PERCTREES	Percent of the lot that is tree covered
	PERCIMP	Percent of the lot that is impervious surface
	PERCGRASS	Percent of the lot that is grass or herbaceous
	PERCBARREN	Percent of the lot that is barren or soil
City database	OWNER	Name of current owner
	TREE_SHRUB	Presence of a tree or shrub
	CITYMAINT	Maintained by the City
	CITYNOTES	Extra information included by the City database; includes indication of new City policy regarding the use of community garden permits
Field data collection using vacant lot assessment tool	OBSERVATION ID	Unique ID for observation
	OBSERVER ID	Observer ID or name
	DATE, TIME	Date and time stamp of observation completion
	PHOTO ID, PHOTOCAPTION	ID number for photo of lot, caption for photo
	VIDEOID, VIDEOCAPTION	ID number for video of lot, caption for video
	BLOCKVIDEOID, BLOCKVIDEOCAPTION	ID number for video of block, caption for video
	LAT, LONG	Latitude and longitude of the point location of the observer when s/he completed the assessment
	ADDRESS	Address of the observer when s/he completed the assessment (STREET, CITY, COUNTY, ZIPCODE, FULL ADDRESS)
	CITY_OWNED	Whether the lot appears to be city-owned (indicated by signage)
	OTHERMAINTENANCE	Whether the lot appears to be maintained by another organization or person

Table 3.1 Variables included in database (parcel level) (Continued)

Source	Variable	Description
	OTHERMAINTAINER	Identity or description of the other maintainer
	KREMER_CAT	Kremer actual use category (0 – Unused land – does not appear actively managed; 1 – Private house/private backyard; 2 – Commercial/industrial; 3 – Community garden; 4 – Park (e.g. forested, paved); 5 – Tree cover between residential buildings; 6 – Sports facilities; 7 – Road, roadside pavement, or sidewalk; 8 – Junk yard – more trash than just litter; 9 – Parking lot; 10 – Non-commercial parking; 11 – Other)
	LOT_ASSETS, ASSET_NUM	Presence or absence of each of the assets specified (BENCHES, GREENTREES, FLOWERS, PATHS, MURALS, SCULPTURE, PICNICTABLES, SHELTER, NATIVEPLANTS, PLAYEQUIP, GARBAGECANS, PLAYFIELD, BIKERACK, RESTROOMS, WATERFOUNTAIN, OTHER); includes an "other" category and the ability to write in new assets. ASSET_NUM represents the sum of categories represented on the lot.
	QUALITYMAINTAINED	"How well are the assets or features of the lot maintained?" (1 – Very well maintained; 2 – Maintained; 3 – In disrepair)
	FOODGROWN, FOODTYPE, OTHERFOODTYPE	Whether food is grown on the lot, the type of food grown (1 – Raised beds or row gardening; 2 – Orchard trees; 3 – Berry canes or shrubs (e.g. raspberries); 4 – Composting; 5 – Rainwater catchment; 6 – Municipal water access; 7 – Other)
	LITTERQUANTITY	Amount of litter present on the lot (1 – None; 2 – A little; 3 – A lot)
	LITTERTYPE	Presence or absence of each type of litter specified (FOODLITTER, ALCOHOLLITTER, TOBACCOLITTER, DRUGLITTER, SEXLITTER, LARGEJUNK, ANIMALREFUSE); includes an "other" category and the ability to write in other types
	GRAFFITI, GRAFFITIPHOTOID, GRAFFITIWORDS	Whether graffiti is present, a photo id for a photo of the graffiti, and the message conveyed by the graffiti ("What does the graffiti say?")
	ABANDONEDBLDGS	How many of the surrounding 6 structures appear abandoned or uninhabitable? (0–6)
	STRENGTHEN, HOWSTRENGTHEN, NEWUSE	Does the vacant lot strengthen the neighborhood? (yes/no); if yes, how does the lot strengthen the neighborhood? (text response); if no, how could the lot be improved to strengthen the neighborhood? (text response)

Property Record (MPROP) database, the City of Milwaukee vacant lot mainte-nance database, the EPA EnviroAtlas, and the field data collected by the project team (Brown, 2014; City of Milwaukee, 2014a; Environmental Protection Agency, 2014). The database makes possible a vast array of potential analyses. For the purposes of this chapter, analyses focused on several key objectives: (1) assess the performance of measures included in the instrument through a compar-ison of the results from two assessors; (2) characterize vacant lots in the Harambee neighborhood; (3) compare characteristics between public and private lots; and (4) provide preliminary information to guide future vacant lot assess-ment driven by a public participatory GIS framework. Descriptive and bivariate statistical analyses were undertaken to provide an understanding of the database and compare key characteristics of interest. Cohen's Kappa statistic was used to assess inter-assessor reliability for selected questions such as litter quantity and the assessor strengthening rating). Quantities of interest (presence of lot assets, assessor strengthening rating) were mapped and explored visually to illustrate the spatial dimension of the data collected. Finally, a logistic regression analysis identified characteristics of lots that predicted an assessor's odds of rating the lot as "strengthening the neighborhood." Independent variables examined in a single, multivariable model included: a sum of the number of assets identified on the lot by the two assessors, public/private status, a sum of the litter rating given to the lot by the two assessors, a sum of the number of abandoned buildings iden-tified, the total land value, the total area of the lot, and the percentage of tree canopy on the lot.

Results

In reference to our previously stated goals, Table 3.2 provides a summary of selected features identified by the assessors for the sample of lots assessed, and the correspondence between assessors. Most of the lots were identified (using the Kremer tool) by both assessors as unused. Gardens (n = 2) and parking lots (n = 2) were identified identically by the two assessors, as were two of what may be three parks. In practice, particular categories presented difficulty, including sepa-rating private use (e.g. a side yard) from unused land, and separating private land from land with commercial use. Our assessors did not identify any of the follow-ing uses on our sample of lots: 5 – Tree cover between residential buildings; 6 – Sports facilities; 7 – Road, roadside pavement, or sidewalk; 8 – Junk yard – more trash than just litter; 10 – Non-commercial parking. One major point of departure for the assessors was on the presence or absence of "trees and greenspace." The City of Milwaukee policy for vacant lots is to plant turf grass to stabilize the space; this stabilization policy was interpreted by only one assessor as "greens-pace." Future assessments should separate discrete elements of the green environ-ment to ensure accurate capture of "green" assets on vacant lots. The following features were not identified on any of the lots assessed by either assessor: bike racks, restrooms, water fountain, drug paraphernalia, sex paraphernalia, and animal refuse. In responding to the question of how many buildings of the six

Table 3.2 Descriptive statistics of lot characteristics: a comparison between assessors

	Assessor 1		Assessor 2	
	n	*%*	*n*	*%*
Kremer actual use category				
0 Unused land – does not appear actively managed	43	82.7	38	74.5
1 Private house/private backyard	1	1.9	7	13.7
2 Commercial/industrial	1	1.9	0	0
3 Community garden	2	3.8	2	3.9
4 Park (e.g. forested, paved)	3	5.8	2	3.9
9 Parking lot	2	3.8	2	3.9
Number of abandoned buildings among six closest buildings				
0	26	50.0	24	46.2
1	20	38.5	21	40.4
2	2	3.8	4	7.7
3	4	7.7	3	5.8
LOT ASSETS (n, % present)				
Benches	3	5.8	4	7.7
Trees and greenspace	6	11.5	45	86.5
Flowers	0	0	2	3.8
Paths	1	1.9	3	5.8
Murals	1	1.9	2	3.8
Sculpture	0	0	1	1.9
Picnic tables	1	1.9	1	1.9
Shelter	1	1.9	1	1.9
Native plants	0	0	1	1.9
Play equipment	1	1.9	1	1.9
Garbage cans	0	0	1	1.9
Playing field	1	1.9	0	0
Quantity of visible litter				
No	16	30.8	15	28.8
A little	31	59.6	35	67.3
A lot	5	9.6	2	3.8
LITTER TYPES (n, % present)				
Food litter	36	69.2	36	69.2
Alcohol litter	15	28.8	19	36.5
Tobacco litter	9	17.3	6	11.5
Large junk	2	3.8	2	3.8

closest were abandoned, the assessors responded similarly. The results of this survey question suggest substantial variability in the built environment context surrounding vacant lots. Many lots had no abandoned buildings nearby, while some lots were surrounded by up to three vacant structures.

Table 3.3 presents a comparison of public versus private lots for selected variables; only data from Assessor 1 was used. Overall, private lots differed more in their maintenance while city lots were stable in their general state. Private lots did not have as many added assets and were commensurately less improved than the public spaces. Private lots also had higher rates of litter and, on average, a greater

Table 3.3 Bivariate analyses of differences between public and private lots (data from Assessor 1)

	Public No.	Public %	Private No.	Private %
Land assessed as having value (P-value: 0.000)				
No	39	97.50	0	0.00
Yes	1	2.50	12	100.00
The current zoning of the property (P-value: 0.162)				
LB2 (commercial zoning)	6	15.00	1	8.30
RM4 (high-density residential)	0	0.00	1	8.30
RT4 (two-family residential)	34	85.00	10	83.30
Mean percent tree canopy (P-value: 0.475)				
(mean, SD)	38.64	26.20	32.55	23.95
Kremer actual use category (P-value: 0.311)				
0 Unused land – does not appear actively managed	33	82.50	10	83.30
1 Private house/private backyard	1	2.50	0	0.00
2 Commercial/industrial	0	0.00	1	8.30
3 Community garden	2	5.00	0	0.00
4 Park (e.g. forested, paved)	3	7.50	0	0.00
9 Parking lot	1	2.50	1	8.30
Number of abandoned buildings among six closest buildings (P-value: 0.560)				
0	19	47.50	7	58.30
1	15	37.50	5	41.70
2	2	5.00	0	0.00
3	4	10.00	0	0.00
Opinion of whether the lot strengthens the neighborhood (P-value: 0.919)				
No	32	82.10	10	83.30
Yes	7	17.90	2	16.70
LOT ASSETS (n, % present)				
Benches (P-value: 0.328)	3	7.50	0	0.00
Trees and greenspace (P-value: 0.526)	4	10.00	2	16.70
Paths (P-value: 0.580)	1	2.50	0	0.00
Murals (P-value: 0.580)	1	2.50	0	0.00
Picnic tables (P-value: 0.580)	1	2.50	0	0.00
Shelter (P-value: 0.580)	1	2.50	0	0.00
Play equipment (P-value: 0.580)	1	2.50	0	0.00
A playing field (P-value: 0.580)	1	2.50	0	0.00
Quantity of visible litter (chi-square P-value: 0.044)				
No	11	27.50	5	41.70
A little	27	67.50	4	33.30
A lot	2	5.00	3	25.00

quantity of litter. The difference between "a little" and "a lot" of litter on public and private lots was statistically significant with a p-value of 0.040 from Fisher's exact test. Private lots were much less likely to appear to strengthen a neighborhood than public lots. The variable assessing litter quantity was evaluated for inter-assessor reliability and performed well, with a Kappa value of 0.70 and 84.6 percent agreement.

A final question on our audit tool asked whether the lot assessed strengthened the neighborhood. If the assessor answered that the lot did strengthen the neighborhood, s/he was asked to describe how it strengthened the neighborhood. If the assessor answered that the lot did not strengthen the neighborhood, s/he was asked how the lot could strengthen the neighborhood. This question was included as a precursor to future participatory GIS efforts that might enhance the community relevance of vacant lot redevelopment projects. Interestingly, the two raters reached a Kappa value of 0.54, with 84.3 percent agreement. Rater 1 noted that nine of the lots strengthened the neighborhood, while Rater 2 reported that 13 lots strengthened the neighborhood. Figure 3.2 presents a visual record of two of the assessed lots. Table 3.4 presents a selection of the dialogue between assessors regarding selection of lots. The range of options considered by both assessors differed significantly. Not only do we expect that different assessors have different priorities concerning the neighborhood needs, but they may be suggesting possible conversions with different timescales and levels of available resources in mind.

This debate among different assessors would have considerable value within a participatory GIS framework, as it would seek to involve neighborhood voices both as sources of local knowledge and as invested agents of the planning and implementation of vacant lot conversion. In the piloting of this assessment tool, the final two qualitative questions were not intended to provide any answers about what should happen to any vacant lots in the current sample, but instead to

Figure 3.2 Examples of vacant lots in Milwaukee, WI, 2014

Table 3.4 A dialogue between raters regarding a selection of vacant lots

Owner	Value	Area	Zoning	Kremer use category	Other (non-City) maintenance?	Current lot assets	Does the lot strengthen the neighborhood?
City	5700	6370	RT4	Unused		1: None 2: Trees and greenspace	1: (N) Use for commercial/residential 2: (N) Brush on edges cleaned to make more open green space
City	0	5664	LB2	Park/unused		1. Playing fields 2. Trees and greenspace	1: (N) If going to be a tot lot needs to be cleaned up made more kid friendly 2: (N) Useable
City	0	1167	RT4	Unused/private	Unclear/owner?	1: None 2: Trees and greenspace	1: (N) Incorporated into a yard 2: (Y) Clean, green
City	0	4780	RT4	Community garden	Milwaukee urban gardens	1: Murals, picnic tables, benches or seating, sheltered area 2: Murals, sculpture, picnic tables, benches or seating, sheltered area, flowers, native plants, trees and greenspace, paths, rain barrels, raised beds or row gardening, composting, rainwater catchment	1: (Y) Community space cared for and provide food and compost 2: (Y) Art, food, community garden, etc
City	0	2264	RT4	Unused		None	1: (N) Use for small local farm 2: (Y) Next to a few other lots so decent open green space
Private	4900	5000	RT4	Unused	Neighbor appears to rake the space/neighbor	None	1: (N) Active backyard 2: (Y) Three in a row so significant green space

test the performance of the questions for future participatory GIS projects with community residents. We considered engagement, debate, and the sharing of information to be valuable processes for the transformation of vacant lots into neighborhood assets, and a set of questions such as the ones trialed here seems well suited to achieve these outcomes. A participatory GIS process may help avoid conversions that lack local buy-in and support and instead provide health assets that are valued and subsequently integrated into the social and physical fabric of the neighborhood.

The number of asset categories identified on the assessed lots by Assessor 1 is represented in Figure 3.3. Although this is a small (10 percent) sample of vacant parcels in the Harambee neighborhood, a pattern begins to emerge that illustrates the presence of more assets on lots in the southeastern corner of the neighborhood. Interestingly, this pattern is similar to that observed by mapping the sum of the assessors' "strengthen" ratings, indicating that assessors in this evaluation viewed the presence of assets as strong contributors to whether a lot strengthened a neighborhood. Further, a logistic regression analysis revealed that the key (and only statistically significant at alpha = 0.05) factors explaining assessors' "strengthen" ratings were the number of assets (sum of two assessor's ratings) (OR = 22.9, 95 percent CI (1.88, 280.09)), and the amount of litter (sum of two assessor's ratings) (OR = 0.05, 95 percent CI (0.01, 0.33)). Future uses of this type of spatial information could include targeting redevelopment efforts to increase the equity of exposure to assets among neighborhood residents.

Discussion

The need for more and better information

Our investigation sought to better understand the salutogenic possibilities of vacant lots in urban neighborhoods. In particular, we were interested in creating a generalizable assessment tool that could provide both qualitative insights and spatial information about vacant lots, their contexts, and their characteristics. In addition to soliciting local knowledge and preferences, the tool's ability to record spatial data and connect with existing geographic resources expands the possible avenues for future analysis. For example, the dosage effect of a vacant lot conversion on a resident could be measured and explored relative to how far that resident lives from the lot. In other words, does proximity matter when it comes to the lot's effect on health? Insights such as these have value in determining how limited resources can be deployed for the greatest possible impact, but these insights are only available with a place-based understanding of health. A spatial analysis of health situates an individual's health firmly within his or her social and physical context.

Furthermore, the spatial data may yield additional insights when connected to other data sources such as localized disease burdens and population health characteristics. Additional neighborhood features that function as health assets or

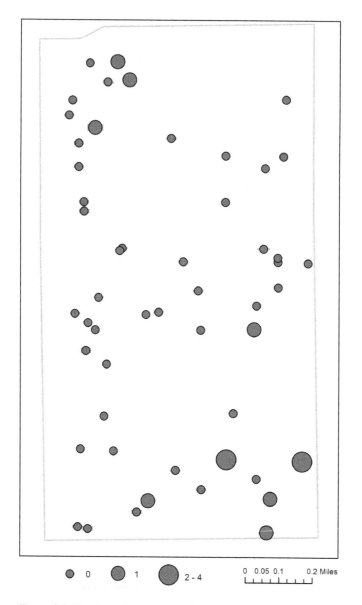

Figure 3.3 Number of asset categories represented on vacant parcels in the Harambee neighborhood, Milwaukee, WI 2014

liabilities, like grocery stores or unsafe traffic patterns, may interact with vacant lots. Significant population health improvement could be realized when vacant lots that weaken a neighborhood are transformed into community gardens, fruit tree parks, play spaces, and other yet unimagined possibilities. Following such changes over time could provide quasi-experimental perspectives on population health that better reveal causal pathways between place and health outcomes. The strategic redevelopment of vacant lots to address public health issues is still a burgeoning area of research, and exactly what characteristics and spatial patterns are most important for potential conversions and activations of the space will continue to require further attention. In ground-truthing our initial tool, and then subsequently assessing a larger sample with an updated audit tool, we have begun to identify some features relevant to the perception of the lot in the neighborhood, such as how well the lot appears to be cared for and the inhabitability of the buildings that surround the lot. We acknowledge that the potential for this tool to be used in a public participatory GIS (PPGIS) framework is as yet untested, but we are hopeful that the tool and its uses may evolve to become increasingly community-informed and of greater usefulness (Dunn, 2007; Knigge and Cope, 2009; Cope and Elwood, 2009).

Defining, locating, and assessing vacant lots can be approached from multiple perspectives, and our initial pilot testing represents only one perspective. Community members residing in neighborhoods with a high concentration of vacant lots can provide rich information about the current use of lots and also be valuable partners in redevelopment efforts. We anticipate that qualitative information will greatly inform the evolution of the vacant lot assessment and potential vacant lot projects. However, solely qualitative feedback cannot elicit an objective assessment of spatial patterns of vacant lots or changes in these patterns over time. Quantitative spatial data is critical to identifying broader patterns of how, when, and why vacant lots are created, maintained, abandoned, transformed, and activated. This data can inform a deeper understanding of the role vacant lots play, and can play, in addressing neighborhood disparities that are co-located with high densities of vacant lots.

Currently, most of the data available for vacant lots is derived from the City of Milwaukee property database. The City of Milwaukee's geographic data represents a uniquely accessible resource for research that may not exist in all cities. Still, no comprehensive and detailed data on the transformation and activation of vacant lots exists in Milwaukee. Without such a resource, the valuable collaboration and cooperation efforts among organizations invested in redevelopment efforts are under-supported and their effects are largely unmeasured. There is a need for an easily accessible, reliable, and standardized method to track vacant lots and their conversions. Community-level work with vacant lot transformations is ongoing, and all parties would be well served to know which conversions work and why they work. Our research mobilized GIS technology to begin to meet this need for data and data-sharing capacity for researchers, City employees, and community organizations and residents. We hope that this GIS vacant lot assessment tool, adapted as needed, will be used in Milwaukee and beyond to engage communities in vacant lot redevelopment efforts, create partnerships between academicians, the City, and the community, and facilitate a central repository for vacant lot data.

There's (not?) an app for that

The opportunity to utilize GIS technology brings forward all kinds of tantalizing possibilities. The embedded geocoding of survey data can reduce the burden of collection, minimize data entry, and help ensure data accuracy. Mapping and organizing data spatially presents great opportunities for sharing findings about vacant lots without requiring the audience to have any specialized knowledge. At the same time, our excitement is tempered by the sometimes unwieldy nature of GIS, particularly in the field. In designing an assessment of vacant lots and choosing a mobile application to carry that assessment, we sought to balance the collection of a rich data set with an approachable and usable mobile GIS data collection application. We found no such "app" that met all of our needs within the commercial market, particularly when considering cost. For this study, we ultimately elected to use the commercial application Fulcrum. Fulcrum proved to be user friendly, allowed for complex survey development, and facilitated offline data collection. Unfortunately, Fulcrum has a commercial model that charges a monthly fee based on the number of users, making it prohibitively expensive as a participatory GIS application.

Additionally, we sought to design an assessment tool and couple it with an application that will effectively solicit as much community input as possible in the future. This means lowering usage barriers from the initial hardware required all the way through the completion of the survey. We conceptualized four key considerations to facilitate assessment completion. Firstly, the application must be generally compatible with mobile computing devices, including phones, and not only with the newest or most expensive models. Secondly, the application must be desirable, meaning that individuals must feel compelled to take part in the project and acquire the necessary app. Thirdly, the data burden of running the application must not preclude its usage; applications that can be utilized offline with cached maps may be preferable given the expense of data plans associated with smart phone technology. Finally, the assessment itself should be user friendly and not overly burdensome so that as many surveys as possible can be completed. If we fail to address these barriers, we fear the exclusion of large segments of potential users and participants and subsequent limitations on the diversity of results and the creation of the best possible vacant lot strategies.

We consider mobile applications to be the most promising route for accumulating neighborhood input, but other options must be made available to facilitate the participation of those unable to access mobile technology. Both an online PPGIS system and offline options for submitting input should be considered resources that complement mobile applications. These alternatives are especially significant when considering that the upfront costs of mobile applications can be thousands of dollars. Costly and often accompanied by user restrictions, mobile apps may disproportionately serve the agendas of funded researchers while community organizations with limited funding are unable to use this technology to promote their priorities.

A future direction for this participatory GIS work could include collaboration with application developers to create a mobile, geographically enabled app specific to the task of gathering vacant lot data that responds to the particular needs outlined above. Few applications have user-friendly GIS services and the ability to support complex conditional surveys, both of which are critical features for our specified purpose. This suggests a disconnect between commercial application designers and users involved in geographical population health research. This gap could be productively filled by geographers, computer scientists, and others, creating productive cross-disciplinary projects with community applicability.

A well-designed and accessible mobile application holds promise as mobile computing continues to grow, but no matter how efficient, the app will not matter if it is not adopted and used by community members. As cities such as Milwaukee consider transforming a surfeit of vacant lots into resources other than traditional development, public participation is crucial for creating lot transformations that are valued and utilized, and yet there are barriers to public participation beyond the technology. Residents may not have the time, resources, or agency to participate in vacant lot projects. Institutions such as the city and established non-profits may create a crowding-out effect whereby their presence mutes the unencumbered dialogue of others. For community engagement and PPGIS to yield positive effects, norms of genuine partnership and equitable engagement need to be cultivated, especially if past experiences have taught residents that community engagement is an empty principle. Academics, city officials, community organizers, and others from positions of power bear responsibility for creating and holding space for local perspectives. Residents must feel welcomed to that space to contribute and feel confident that their knowledge and values will drive the process and shape the results in a meaningfully way. When norms match actions and adequate resources are made available, then PPGIS and other participatory methods create a sense of local ownership of the new vacant lot assets.

Furthermore, we argue that vacant lot redevelopment without first partnering with the community runs the risk of creating an undesirable and unsustainable conversion that may not have the intended positive effect. Even if the conversion is objectively health-promoting, a conversion will only find sustainable success with neighborhood support. In a community where neighbors are able to garden in their own backyards and are uninterested in a community garden, such a lot conversion may have no tangible health benefit and even regress to an uncared for and unrealized neighborhood space. By contrast, an outdoor play space that neighbors sought and valued may have flourished in the same space.

The strategic development of community partnerships is a critical step for mobilizing GIS technology for participatory purposes. The value and functionality of GIS applications may not be apparent to users who are not familiar with this technology and some foundational cooperative work is called for. We anticipate this early investment will better integrate local knowledge with the institutional needs and requirements of the vacant lot's current owner – the city. PPGIS can translate anecdotal local understandings into an aggregated spatial data format that is more likely to inform the work of city officials simply because that knowledge is now readily available and interpretable. In order to implement a PPGIS process, relationships with community organizations need to be developed, collaborative planning initiated, and relevant training undertaken. Specifically, community organizations can provide great value in helping identify relevant data to collect and in soliciting volunteered geographic information and project participation more broadly within neighborhoods. Engaging community members as co-researchers along the continuum of the project increases the likelihood that the community will drive the research process and value its outcomes. Similarly, when the design and implementation of population health improvements is guided by community participation, it can create more lasting, sustainable, and effective neighborhood modifications.

Findings and futures

As anticipated, the vacant lot assessment yielded a range of vacant lot types including converted spaces like community gardens and private vacant lots that appeared to be untended dumping grounds. The City of Milwaukee policy to stabilize city-owned vacant parcels was clearly observed over the course of the assessment. The city-owned lots were cared for spaces of mown turf grass, and nearly all city-owned lots were clearly identified as such with consistent signage. City-owned lots often contained litter but so did other public spaces within the Harambee neighborhood such as sidewalks and medians. City-owned lots were significantly less likely to have "a lot" of litter compared to the private lots within the sample. Relatedly, the absence of litter and the presence of assets correlated with the likelihood of the assessor identifying the lot as strengthening the neighborhood. The presence of multiple lot assets, such as benches, increased the likelihood that the assessor would identify the lot as strengthening the neighborhood while the presence of litter, a little or a lot, increased the likelihood of a "weakens the neighborhood" assessment. Anecdotally, the assessors found that vacant lot conversions were proximate to schools, homes in good repair, or other preexisting community assets. We anticipate that larger samples will reveal additional spatial patterns and may bear out anecdotal assessments such as this.

These observations may be interpreted in multiple ways. For example, vacant lot improvement efforts around schools or well-maintained houses may indicate the mobilization of resources that are not available on blocks where a greater number of homes are abandoned or in disrepair. Alternately, improvement efforts near schools may indicate the value the community places on educational institutions and a strategic investment in areas where children congregate. If positive vacant lot transformations are clustered in areas of relative resource wealth, partnerships and outside resources may be appropriate approaches to enabling the revitalization of areas where residents are unable to act as independent investors.

It is also possible that blocks that house buildings in disrepair may be used for illicit or illegal behavior, posing a safety risk for neighbors otherwise interested in redeveloping that area. Given the breadth of possible contributors to these patterns of uneven vacant lot redevelopment, engaging community members with local expertise is essential for improving the vacant lot assessment tool and better understanding what vacant lot patterns mean for neighbors' day-to-day experiences.

Emerging redevelopment strategies are moving away from the presumption that vacant lots will house residential properties again. With this shift towards a diverse array of vacant lot possibilities, including but not limited to residential use, the importance of community buy-in to enable sustainable, context-appropriate use of these spaces is even greater. Interventions that address the diverse issues found in these neighborhoods – lack of green space, atrophied economies, limited fresh food access, absence of health and community centers, and more – are critical. Community ideas, what is feasible on the lot, and the development resources available all inform the intervention design and implementation. Cities can facilitate new urban patterns and paradigms of development that serve residents' needs by promoting innovation and experimentation. By revisiting and revising permit, tax, and zoning structures and creating incentives for development that provides accessible jobs and health-promoting goods such as better access to food or health services, cities can concurrently decrease the number of vacant lots and reduce neighborhood disparities. As cities, organizations, researchers, and community members explore new avenues of intervention and redevelopment, the success of vacant lot transformations will depend on a transparent, accessible process that values multiple perspectives on cultivating population health. With cooperation and collaboration, more can be done than just mitigating the deleterious symptoms of vacant spaces. Partnerships can catalyze and sustain productive and valued spaces that promote the health and well-being of both neighbors and neighborhoods.

4 Spatial analysis of HIV/AIDS survival in Dallas County, Texas

Joseph R. Oppong, Stephanie Heald,
Warankana Ruckthongsook, and
Chetan Tiwari

Due to significant advances in testing, prevention, and treatment, individuals with HIV have much higher survival rates and life expectancy than in the past. Early diagnosis of HIV infection provides timely access to highly active antiretroviral therapies (HAART), and improves patient outcomes, enabling individuals with access to this treatment to live much longer and fuller lives (Palella *et al.*, 2003; Kitahata *et al.*, 2009). In contrast, late diagnosis means poorer health outcomes, increased mortality risk, and missed opportunities to reduce onward transmission, for example through reduction of high-risk behaviours (Egger *et al.*, 2002; Marks *et al.*, 2006). Survival disparities between those with access to HAART and those without are well documented (Poznansky, 1995; Katz *et al.*, 1998; Rapiti *et al.*, 2000; Montgomery *et al.*, 2002; Borrell *et al.*, 2006; Walensky *et al.*, 2006; Harrison *et al.*, 2010).

HIV infection advances in three stages – acute, chronic, and acquired immunodeficiency syndrome (AIDS). Stage 1, acute HIV infection, generally develops within two to four weeks after a person is infected with HIV. At this stage, HIV multiplies rapidly and spreads throughout the body, attacking and destroying the infection-fighting CD4 cells of the immune system, and the risk of transmission to others is greatest. In Stage 2, chronic HIV infection, the HIV virus continues to reproduce at very low levels, usually with very mild or no symptoms. Without treatment, chronic HIV infection, Stage 2, usually advances to Stage 3 (AIDS) in ten to twelve years. Treatment with HAART delays and can even prevent progression to Stage 3 (AIDS). Without treatment, people who progress to AIDS typically survive about three years. Progression through these stages varies among people depending on several factors, particularly rapid, sustained access to treatment (Hall *et al.*, 2006). Thus early diagnosis and treatment are critical determinants of HIV survival. Late testers have poorer survival because the virus has already done considerable damage by the time of diagnosis.

In the United States, HIV prevalence varies widely among the states, and Texas, the focus of this study, has the fourth highest number of people living with HIV, following New York, California, and Florida (Centers for Disease Control (CDC), 2014). In addition, survival for people living with HIV/AIDS varies geographically, between genders, and among race/ethnic groups (Singh *et al.*, 2013). For example, women survive longer than men, and Black males, followed by Hispanic males, have the lowest survival rates (Harrison *et al*, 2010). Similarly, the US South region reported the lowest survival between 2003 and 2007 (CDC, 2013a; Oppong *et al.*, 2014).

In Texas, HIV survival varies significantly between race/ethnic groups (Texas Department of State Health Services (TDSHS), 2013), but the geographic pattern and drivers of disparities in survival remain unclear. This is the knowledge gap we seek to fill. Using zip code-level analysis, we examine the geography of HIV/ AIDS survival measured by Kaplan-Meier analysis. For all cases, we define HIV survival as the time in months from the date of HIV diagnosis to the date of Stage 3 (AIDS) diagnosis (CDC, 2014). For late testers alone, we also examined the time from HIV diagnosis to death.

Determinants of spatial patterns of survival

Numerous studies confirm that geographic location and socioeconomic conditions affect health outcomes (Wallace, 2003; Fiscella and Williams, 2004; LaVeist, 2005), and that health status varies significantly between places. Within the urban realm, low socioeconomic status (SES) has been linked to poorer health outcomes and higher mortality rates (Fiscella and Williams, 2004; LaVeist, 2005). Furthermore, communities of low socioeconomic status tend to be located in close proximity to other disadvantaged communities, thus resulting in a concentration of poverty and disadvantage that makes these places vulnerable to disease (Oppong and Harold, 2009). More broadly, place vulnerability theory argues that adverse life circumstances do not affect all places uniformly, and that certain factors make some places more vulnerable to diseases and death than others (Oppong and Harold, 2009). Thus, mapping and analyzing the spatial distribution and associated characteristics of these vulnerable areas can help to explain some of the disparities that exist in HIV/AIDS survival in urban Texas and to identify places and populations that most urgently need public health interventions.

In both Texas and nationally, Blacks face the most severe burden of HIV/ AIDS. The rate of HIV infection among Blacks is four times higher than the rate for white and Hispanic Texans (TDSHS, 2009). Blacks are also the group most likely to reside in large urban communities with concentrated poverty (Sanez, 2005). Previous studies suggest that race/ethnicity can be an important risk marker for poor health outcomes and shortened survival (Fiscella and Williams, 2004; Hall *et al.*, 2006). Socio-economically disadvantaged areas, often over-crowded and characterized by substandard housing, facilitate rapid disease spread and may serve as incubators for disease or breeding grounds for opportunistic infections (Drucker *et al.*, 1994; Wallace and Wallace, 1997; Galea and Vlahov, 2005). Opportunistic diseases such as tuberculosis, meningitis, bacterial pneumonia, encephalitis, and specific cancers can attack the weakened immune systems of HIV/AIDS patients and shorten their lives considerably (UNAIDS, 1998). Groups vulnerable to HIV/AIDS may also be more vulnerable to shortened survival due to the likelihood of contracting communicable opportunistic diseases. In such vulnerable places, the likelihood of individuals engaging with other individuals with compromised immune systems is also substantially higher (Latkin *et al.*, 1995; Geronimus, 2000). Thus the socio-economic characteristics of the place of residence of a person living with HIV can increase their vulnerability to specific communicable diseases, resulting in a decrease in their survival.

Socio-economic status is a barrier to receiving the adequate medical care needed for improving survival chances for an individual with HIV. Due to cost, only 57 percent of HIV patients nationwide who are aware of their infection are estimated to be in care (Walensky *et al.*, 2006; Schackman *et al.*, 2006). Numerous studies have concluded that individuals who lack health insurance, even if only for a short period of time, have significantly higher mortality rates than those who are insured (Hadley, 2003; Hadley *et al.*, 2008; Ayanian *et al.*, 2000). Moreover, minorities make up a large portion of uninsured adults and tend to have less access to medical services (Brown *et al.*, 2000). Previous studies, using hazard models to calculate HIV/AIDS survival, demonstrate that it varies among socio-economic groups (Katz *et al.*, 1998; Arnold *et al.*, 2009; Meditz *et al.*, 2011).

In addition to the socio-economic and physical characteristics of a neighborhood and financial restrictions on access to healthcare, late testing may impact HIV survival (Castilla *et al.*, 2002). Late testing, or being diagnosed with Stage 3 (AIDS) within one year of an HIV diagnosis, delays the initial treatment of an HIV-infected individual and is known to have a significantly negative impact on disease prognosis (Poznansky *et al.*, 1995; Kitahata *et al.*, 2009; US Department of Health and Human Services, 2013). Late testers have significantly shorter survival with HIV/AIDS than non-late testers (Hocking *et al.*, 2000; Castilla *et al.*, 2002; Girardi *et al.*, 2004). Early testing is also vital for reducing the overall burdens of HIV in a population, as it prompts infected individuals to make behavioral changes that can help prevent the spread of the disease to others (Crepaz *et al.*, 2009). As of 2009, almost one-third of those living with HIV in Texas were late testers (TDSHS, 2009).

This study investigates the spatial variation of HIV/AIDS survival and explains its geographic distribution using neighborhood characteristics including measures of income, poverty, education, and unemployment. Specifically, we analyze the social and geographic inequalities in HIV/AIDS survival in Dallas County, Texas. We examine associations at the zip code scale between survival and socio-economic status (SES), minority population concentration, and rates of late testing for HIV.

Identifying spatial patterns of HIV/AIDS survival is imperative for targeted intervention. Knowing the locations and characteristics of populations that are susceptible to poorer survival enables us to identify service gaps, determine appropriate spatial intervention strategies, and evaluate intervention programs aiming to increase survival and reduce disease burden.

Study area

In the United States, individuals living in urban areas are more vulnerable to contracting HIV/AIDS. In 2011, large metropolitan areas had the highest HIV prevalence rates (431.8 per 100,000 population), which is three times higher than in non-metropolitan areas (128.5 per 100,000 population) (CDC, 2013b). For this reason, much of the HIV/AIDS research in the US has targeted large cities such as New York and San Francisco (Katz *et al.*, 1998; Lee *et al.*, 2001; Wallace, 2003; Walensky *et al.*, 2006; Harrison *et al.*, 2010). However, because few large cities in the South have been the focus of HIV/AIDS survival analyses, Texas, with over 32,000 HIV/AIDS cases, is a perfect setting for this study (CDC, 2012). We focus on Dallas County, a metropolitan area (Dallas–Fort Worth Metroplex) with densely crowded urban cores and neighborhoods, varying socio-economic characteristics, and the second largest concentration of people living with HIV/AIDS in Texas (CDC, 2012).

Only the 82 zip codes that have their centroids within the county were selected for inclusion in the study (Figure 4.1). Because zip code boundaries are created by the postal service, and county boundaries are political entities created by the state, boundaries of the two spatial units do not perfectly match.

Data and methodology

We used HIV/AIDS data obtained from the Texas Department of State Health Services. The data set contained an anonymized list of all HIV/AIDS cases at the zip code level between January 1999 and December 2008. The data included zip code of residence at diagnosis, gender, race/ethnicity, and diagnosis date. Dallas County had 12,369 cases in total. Of this number, 490 cases, confirmed dead but without a date of Stage 3 (AIDS) diagnosis, were removed from the analysis. The remaining 11,879 records were geocoded to the Zip Code Tabulation Areas (ZCTAs) using ArcGIS 10.2 and cartographic boundary files obtained from the 2010 US Census. After geocoding, an additional four records were removed due to unmatched zip codes. The remaining 11,875 cases (96 percent of total cases), comprising 5,890 HIV and 5,985 confirmed AIDS cases, were included in the analysis.

For neighborhood characteristics, we used 38 socio-economic variables, recommended by the CDC as indicators of poverty. The variables included measures of educational attainment such as education through the 8th grade and through doctorate degree, median family income, median household income, labor force characteristics including unemployment rate, and employment in specific sectors such as sales, management, service, construction, farming, and production. These variables were derived from the 2010 US Census Bureau. All measures for each zip code within the study area were entered into a Principal Components Factor Analysis to create composite variables describing socio-economic characteristics of the zip code population.

Factor analysis is commonly used to reduce large numbers of variables into smaller numbers of identifiable dimensions (Haan and Kaplan, 1985; Vyas and Kumaranayake, 2006). Highly correlated variables, whether positive or negative, are basically influenced by the same factors and are combined into an identifiable dimension known as a factor. From the 38 original variables, we extracted three

Figure 4.1 Dallas County, Texas

dimensions that we characterized as extreme low, moderate, and high socio-economic status (SES) (Table 4.1). These three factors explained approximately 86 percent of the shared variance of the 38 variables. The high SES dimension corre-lated positively with higher levels of educational achievement, high median family

Table 4.1 Highest significant factor loadings of the 38 selected variables included in the factor analysis which identified three distinct groups

Variables with high factor loadings	Socio-economic groups		
	Extreme low SES	*Moderate SES*	*High SES*
Median family income 10,000–14,999	*0.934*	0.243	0.095
Median family income 60,000–74,999	0.258	*0.840*	0.412
Median family income 150,000–199,999	–0.012	0.291	*0.853*
Educational achievement: Bachelor's degree	0.114	*0.507*	*0.824*
Educational achievement: 9th grade	*0.915*	0.311	0.016
Unemployed population over 16	*0.809*	0.411	0.244

Note: Italic indicates highest of the high correlations.

income, high median household income, and high employment. The extreme low SES dimension, on the other hand, correlated with low levels of educational attainment, high unemployment, low family income, and low median household income. Moderate SES had characteristics that fell between the high and extreme low SES groups, as evidenced by high concentrations of population with moderate levels of income and educational attainment. Table 4.1 shows some of the highest and most significant factor loadings of the three identified SES groups.

Using Bartlett factor scores we identified SES status for each zip code. Bartlett factor scores are standardized scores similar to a Z-score metric, where values range from –3.0 to +3.0 approximately. Since factor scores are computed based on the composition variables (DiStefano *et al.*, 2009), the factor with the highest score indicated the SES status of that zip code. Next, these zip codes with identified SES status were mapped to illustrate their spatial distribution.

Geographically, the extreme low SES areas were zip codes in or immediately surrounding the urban core of Dallas. The moderate SES areas were mostly zip codes surrounding the extreme low SES zip codes. Further in the periphery, particularly in the north, were areas of high SES (Figure 4.2).

HIV survival – which is defined as the time in months from the date of HIV diagnosis to the date of Stage 3 (AIDS) – was calculated using Kaplan-Meier survival analysis. This measure was computed for each individual case and then averaged for the zip codes characterized by the three distinct SES groups. The term "survival" can be deceptive because techniques are essentially designed to measure an individual's time to an event (Smith and Smith, 2004). However, the data used in this study includes individuals who did not reach an event, Stage 3 (AIDS), before the end of the study period. These individuals were considered as right censored-data for the Kaplan-Meier analysis (Kaplan and Meier, 1958). Then, survival time across SES groups was compared using a log-rank test to determine whether survival distributions vary between SES groups.

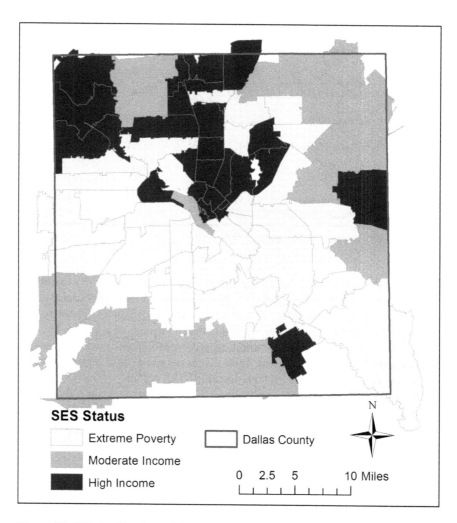

Figure 4.2 SES classifications of the Dallas County zip codes

For the analysis of late testers, the event of HIV survival is defined as time from HIV diagnosis to death; individuals who did not die during the time of study were considered as right censored. We present differences in survival between SES areas, race/ethnic groups within each SES area, and between late and early testers.

Results

The average rate of new HIV infections between 1999 and 2008 for Dallas County is 49.0 per 100,000 population, which is about three times higher than the 2010 average for Texas (17.4 per 100,000 population) (TDSHS, 2011). From 1999 to

Table 4.2 HIV to Stage 3 (AIDS) survival months in Dallas County

SES groups	Median survival months (95% confidence intervals)	Total cases	Event cases
Extreme low SES	52.0 [47.385, 56.615]	6,670	3,524
Moderate SES	46.0 [35.529, 56.471]	1,524	773
High SES	79.0 [68.180, 89.820]	3,681	1,688
Overall	58.0 [53.811, 62.189]	11,875	5,985

2008, 5,985 of 11,875 cases (50.4 percent) had developed to Stage 3 (AIDS) and the overall median survival time, or the time between HIV diagnosis and Stage 3 (AIDS) diagnosis for 50 percent of the population, was 58 months (Table 4.2). When classified by SES groups, the median survival of individuals living in areas considered high SES was 79 months while individuals living in extreme low and moderate SES areas had much lower median survival times at 52 and 46 months respectively (Table 4.2). Interestingly, individuals in moderate SES zip codes had the shortest survival with HIV. Thus, the likelihood over time for HIV cases to reach Stage 3 (AIDS) is higher for those individuals living in areas characterized as moderate SES. The median survival months of the three SES groups were mapped to illustrate the residential locations in which HIV survival is shortest (Figure 4.3).

Figure 4.4 presents Kaplan-Meier plots for the analysis comparing progression of the disease over time for individuals living in each SES area. Although individuals living in moderate SES areas had the shortest median survival with HIV, the Kaplan-Meier plot showed that the progressions of disease over time between individuals living in extreme low and moderate SES areas were slightly different. From the first month of HIV diagnosis until 80 months, the percent of the population progressing to Stage 3 (AIDS) in areas of moderate SES was slightly higher than that in extreme low SES areas. However, after 80 months individuals living in areas of low SES had poorer survival (Figure 4.4). The population living in high SES areas had the longest survival with HIV for the entire period (Figure 4.4).

The log rank test confirmed that the survival distributions for people living in the different SES regions were not equal ($p < 0.05$). The pairwise comparison indicated that the survival distribution of individuals living in high SES areas was significantly higher than that of individuals living in extreme low and moderate SES areas ($p < 0.05$) while there was no statistically significant difference between individuals living in extreme low and moderate SES areas ($p > 0.05$).

Survival within socio-economic status (SES) areas by race/ethnicity

Of the 2.4 million population of Dallas County, Hispanics comprised the largest race/ethnic group, accounting for 38 percent of the total population; non-Hispanic white comprised 33 percent and 22 percent were Black (US Census, 2010). Blacks accounted for the highest proportion of newly diagnosed HIV (47 percent of total HIV/AIDS cases) (Figure 4.5), and the Black HIV rate, 106.3 per 100,000 population, was two and four times higher than White (46.2 per 100,000

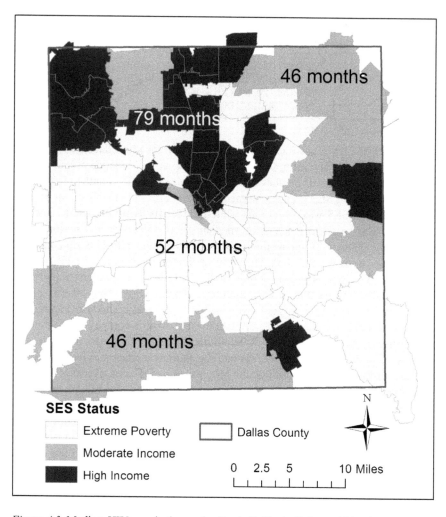

Figure 4.3 Median HIV survival months for individuals living within the three SES
regions

population) and Hispanic (24.6 per 100,000 population) rates respectively.
Interestingly, despite having the lowest HIV rates, Hispanics had the shortest
HIV median survival at 24 months compared to other race/ethnic groups; White
and Black median survival times were estimated to be 105 and 50 months respectively (Table 4.3). Also, a Kaplan-Meier plot of HIV survival by race/ethnic
groups confirms that Hispanics had much shorter survival over time. At the end
of 120 months, while half of the Whites survived, only 30 percent and 35 percent
of Hispanics and Blacks, respectively, did (Figure 4.6(a)). The log rank test

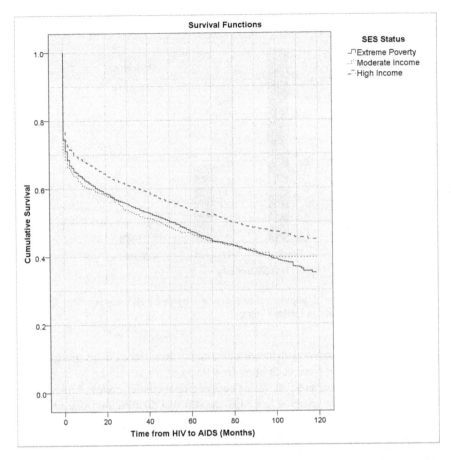

Figure 4.4 Kaplan-Meier HIV survival plot showing the percentage of individuals within each SES group's progression from HIV to Stage 3 (AIDS)

Table 4.3 HIV to Stage 3 (AIDS) survival months by SES and race/ethnic groups

SES groups	Median survival months (95% confidence intervals)		
	Whites	*Blacks*	*Hispanics*
Extreme low SES	101.0 [NA, NA]*	53.0 [47.086, 58.914]	15.0 [8.193, 21.807]
Moderate SES	62.0 [35.992, 88.008]	47.0 [33.301, 60.699]	7.0 [0, 19.812]
High SES	115.0 [NA, NA]*	53.0 [41.455, 64.545]	46.0 [28.530, 63.470]
Overall	105.0 [NA, NA]*	50.0 [45.057, 54.943]	24.0 [17.033, 30.967]

* Confidence intervals of the median survival of HIV among Whites regardless of SES areas and those living in extreme low and high SES areas could not be estimated since their survival times were close to the end of study time (120 months).

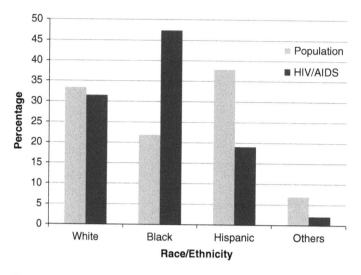

Figure 4.5 Percentages of population and newly diagnosed HIV cases, 1999–2008, by race/ethnicity in Dallas County

confirmed that the survival distributions among these race/ethnic groups were significantly different from each other ($p < 0.05$).

In each SES region, survival varies between race/ethnic groups. Among race/ethnic groups living in extreme low SES, the survival distribution from a Kaplan-Meier plot (Figure 4.6(b)) had a similar pattern as HIV survival by race/ethnicity regardless of SES groups (Figure 4.6(a)), and the statistical test indicated that HIV survival differed significantly ($p < 0.05$) among race/ethnic groups. The median survival of Hispanics was approximately 3.5 and 7 times shorter than that for Blacks and Whites respectively (Table 4.3). In moderate SES areas, the median survival time among Hispanics was only 7 months compared to 47 and 62 months for Blacks and Whites respectively (Table 4.3). Also, the survival distribution of HIV among Hispanics was significantly different from other race/ethnic groups ($p < 0.05$), but there was no statistical difference between Blacks and Whites ($p > 0.05$) (Figure 4.6(c)). For high SES areas, Hispanics had the shortest median survival time (46 months), while the median survival times for Blacks and Whites respectively were 53 and 115 months (Table 4.3). However, a Kaplan-Meier plot of HIV among race/ethnic groups living in high SES region (Figure 4.6(d)) illustrated that the survival distribution of Blacks and Hispanics were close to each other over time, and the pairwise comparison test confirmed that there was no statistical difference between them ($p > 0.05$). Moreover, Figure 4.6(d) showed that the survival distribution of HIV among Whites was much better than for Black and Hispanics over time ($p < 0.05$).

In summary, comparing HIV survival by race/ethnic groups and SES areas together, Hispanics living in all SES areas experienced the worst median survival

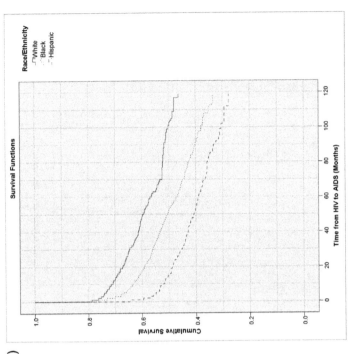

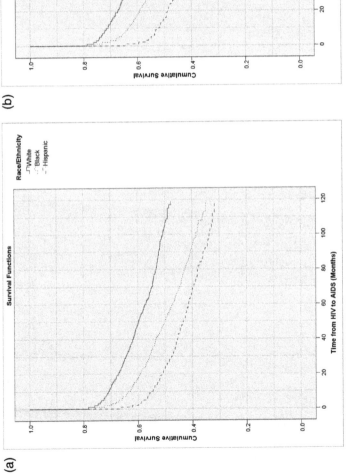

Figure 4.6 Kaplan-Meier HIV survival plot showing progression from HIV to Stage 3 (AIDS) by SES and race/ethnic groups: (a) race/ethnic groups regardless of SES; (b) extreme low SES by race/ethnic groups

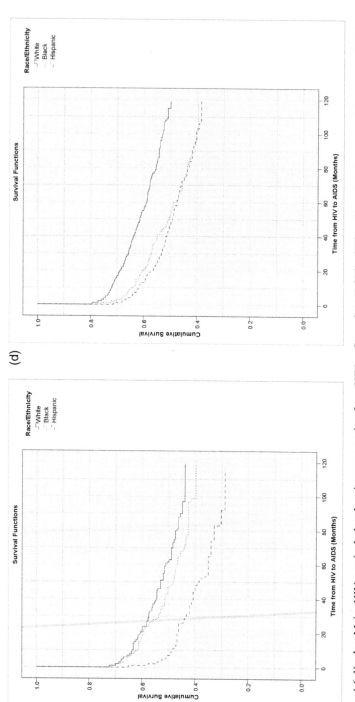

Figure 4.6 Kaplan-Meier HIV survival plot showing progression from HIV to Stage 3 (AIDS) by SES and race/ethnic groups: (c) moderate SES by race/ethnic groups; (d) high SES by race/ethnic groups

time, especially those living in moderate SES areas. Whites had the longest median survival time in all SES areas, particularly individuals living in high SES areas. Also for all race/ethnic groups, those living in moderate SES areas had the poorest survival.

Survival and late testers

In Dallas County, from 1999 to 2008, 4,424 of 11,875 cases (37.3 per cent) were identified as late testers. This is higher than the Texas average of 33.5 percent (Oppong *et al.* 2012). Late testing was highest among those living in moderate SES areas, at 40.3 percent, followed by 38.6 percent and 33.6 percent late testers of those living in extreme low and high SES areas, respectively. These percentages are well above the state average. Additionally, HIV/AIDS survival was much worse for late testers compared to non-late testers ($p < 0.05$). Since less than half of cases reached the event (death), the average survival time was used instead of the median survival time. Late testers on average lived approximately 94 months with HIV/AIDS while non-late testers lived almost 115 months (21 months longer). Also, individuals who were non-late testers had 90 percent survival compared to 70 percent for late testers (Figure 4.7).

Discussion

Our analysis of HIV survival at the zip code level in Dallas County reveals important differences in survival by socio-economic status and among racial/ethnic groups. While high SES areas had significantly higher HIV survival (better health outcomes), HIV survival was shortest for those living in moderate SES zip codes. Also Hispanics had poorer survival than Blacks. This contrasts with previous research, which suggests that areas of poverty and urban decay consistently have poorer health outcomes and Blacks have the worst survival (Harrison *et al.*, 2010).

Although contrary to previous studies, these findings make sense because low SES individuals are more likely to have access to free or reduced-fee HIV care provided by government programs such as Ryan White than individuals with higher incomes. In Texas, populations that fall below the poverty line are eligible for such benefits, thus enabling them to have access to highly active antiretroviral therapy (HAART). For example, the Texas HIV Medication Program (THMP), the government-funded AIDS Drug Assistance Program (ADAP) for Texas, requires proof that the client has exhausted all other resources available to them and has an annual adjusted gross income at or below 200 percent of the current federal poverty income guidelines. In addition, every six months, they must provide proof of continued eligibility – proof of identification, residency, HIV diagnosis, and meeting low-income requirements. These requirements clearly favor extreme low income people but not those with moderate income. Consequently, it is much easier for low-income people, many of whom live in the extreme poor SES areas, to meet and maintain financial eligibility requirements and to stay in treatment, a critical factor in survival (Hall *et al.*, 2006).

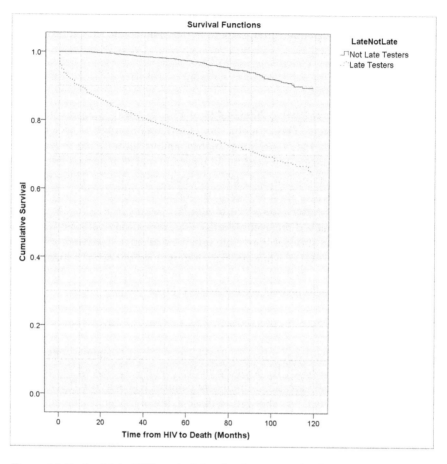

Figure 4.7 Kaplan-Meier HIV survival plot showing progression from HIV/AIDS to death for late testers

Also, people with HIV/AIDS living in extremely low SES zip codes had better access to government assistance programs such as Ryan White that prolong survival with HIV. These people are more likely to receive medical treatment supplemented with ongoing case management services to facilitate continuity of care as well as supportive services such as transportation, legal assistance, nutrition services, mental health services, substance use treatment, and other essentials for living with HIV. For example, the Ryan White program covers many critically important services for low-income people living with HIV/AIDS – non-medical case management, legal services, food bank and home delivered meals, linguistic services, housing services, and emergency financial assistance – that are not available to beneficiaries of Medicaid or other programs (Center for Health Law and Policy Innovation of Harvard Law School and the Treatment Access Expansion Project, 2013). Due to the importance of these services for the well-being of people living with HIV/AIDS, HIV patients using Ryan White program services have higher care-retention rates than HIV patients not in the program (Center for Health Law and Policy Innovation of Harvard Law School and the Treatment Access Expansion Project, 2013). Consequently, the individuals who transition from the Ryan White program on to Medicaid or private insurance plans are likely to be at a disadvantage.

An especially vulnerable population is late testers who not only get sicker, but also are deprived of all the benefits associated with early initiation of treatment (Kitahata *et al.*, 2009). Thus a pressing need exists to intensify prevention efforts and surveillance in communities to reduce late testing and to rapidly link newly diagnosed cases with care. HIV medications not only improve the health of the individual, they also reduce infectiousness, thus decreasing the risk of transmitting HIV to others. An estimated one third of people living with HIV in the United States are not in care (Health Resources and Services Administration, HIV/AIDS Bureau, 2006).

In high SES zip codes, individuals are more likely to be able to afford treatment options either through personal funds or private health insurance. This explains their high survival rates compared to individuals living in other SES areas. Also, this group has the lowest percentages of late testers. Living in high income areas may ensure better access to the best treatment options and, thus, increased survival.

People living in moderate SES zip codes are the most vulnerable to shortened survival. More than likely, the people who reside in these areas are too poor to afford costly treatments (Schackman *et al.*, 2006), but too wealthy to be eligible for Ryan White and Medicaid in the initial stages with HIV/AIDS. Although such individuals may eventually qualify for Medicaid for disability or other reasons, the critical time for initial diagnosis and treatment would have likely passed. Thus affordability and access to treatment remain obstacles for people living with HIV in moderate SES areas. Also in our study area, moderate SES areas recorded much higher percentages of late testers.

We observed significant differences in survival between race/ethnic groups. Hispanics had shorter median HIV survival than both Blacks and Whites. Previous research suggests that minorities, especially Blacks, experience poorer health outcomes and higher mortality overall (CDC, 2016). Guwani and Weech-Maldonado (2004) report that Blacks continue to face barriers in accessing healthcare despite protections under Medicaid. In contrast, our results from the Kaplan-Meier analysis show that Blacks had higher rates of HIV survival than did Hispanics. This may be due to increased access to government-funded treatment programs since in our study, Blacks were the predominant race/ethnic group in extreme low SES zip codes.

Hispanics have the lowest percentages of HIV/AIDS cases in moderate and high SES areas, 14.1 percent and 19.9 percent respectively, while they have similar proportion as Whites, 20 percent approximately, in the extreme low SES areas. Additionally, Hispanics may experience language and other barriers which limit access to healthcare. In our study, most Hispanics live in zip codes characterized as moderate SES. Consequently, as we have argued, they may earn too much to qualify for Medicaid initially. Socially, according to Therrien and Ramirez (2000), it is common for Hispanics to have larger household size, so it is conceivable that opportunistic diseases such as TB or hepatitis could more easily spread in the crowded homes. Lastly, to receive Medicaid benefits, a person must be an American citizen, in the process of naturalization, or have a valid work visa (US Department of Health and Human Services, 2011). Some Hispanics may be living in Texas illegally, and be unable to qualify for Medicaid or afford treatment.

One clear limitation of this work is that we do not have details of individual cases such as income, level of education, or employment status. Thus we are compelled to use an ecological approach and assume that the characteristics of the zip code of residence reflect those of the individual. Teasing out the fine nuances and reasons for differences in survival requires more detailed individual level data which is not available due to confidentiality restrictions.

Conclusion

The spatial pattern of HIV survival is complex and appears to vary with the socio-economic characteristics of zip codes of residence and race/ethnicity. The most vulnerable to poor survival may not be HIV patients in extremely low SES zip codes, but rather those in the moderate SES zip codes who are too rich to benefit from government programs, but too poor to cover the costs of HIV treatment. Our results also suggest the importance of targeting certain vulnerable population groups, in our study Hispanics and Blacks living in moderate income SES zip codes. Also there is an urgent need to examine place characteristics – access to services, transportation barriers, etc. – that might deter people from gaining access to diagnosis and treatment services. Spatially targeted interventions may reduce the rate of late testing, decrease the rate of new infections, and increase the length and quality of life for Texans living with HIV/AIDS.

5 From cultural clashes to settlement stressors

A review of HIV prevention interventions for gay and bisexual immigrant men in North America

Nathaniel M. Lewis

During the past twenty years, gay men's health practitioners in North America have increasingly eschewed "one size fits all" sexual health and HIV prevention interventions for gay men in favor of those tailored to gay men and men who have sex with men (MSM) of specific age and ethno-racial backgrounds. Accordingly, health researchers, agencies, and clinics have tested and evaluated dozens of interventions targeted toward Latino, Asian, and African-American/Afro-Caribbean men, many of whom are also immigrants to the United States and Canada. Yet most of these interventions, which have included retreats, peer counseling programs and sexual health information sessions, seek to address or employ elements of "culture" and "acculturation." Typically, it is the immigrant's ethnic identity or culture, not migration itself, which is thought to inform sexual risk-taking behaviors or related issues such as substance use. Consequently, interventions in the gay/MSM immigrant population alternately challenge elements of cultural orientation perceived as destructive (e.g. internalized homophobia) and build on those perceived as protective (e.g. family support and respect). Relatively few, however, directly address the role of *the migration experience* – including stress, isolation, and encounters with new communities – that affect the context of sexual risk. This chapter therefore seeks to enumerate and categorize North American sexual health and HIV prevention interventions employed in (but not always targeted to) immigrant MSM communities, their intended effect, their success and the degree to which they address elements of migration and resettlement versus culture and acculturation.

At the height of the AIDS crisis during the 1980s and 1990s, the health research community began developing interventions to stem rampant HIV infection among gay men and MSM. While effective, these interventions (e.g. Kelly *et al.*, 1991, 1997) tended to recruit mostly urban, white, middle-class men from mainstream "gay community" events and venues such as bars, bath houses, and Pride festivals. Later interventions expanded in scope to target, for example, Black men (e.g. Wilton *et al.*, 2009), rurally situated men (e.g. Bowen *et al.*, 2006), and young men (e.g. Bull *et al.*, 2009). Interventions targeting diverse ethnic groups of gay men/MSM, particularly those comprising mostly foreign-born men, have been sparse. This gap has been attributed to the high cost of intervention research and the lack of researchers and trainees who represent minority and immigrant groups (see Ramirez-Valles, 2007). At the same time, groups such as Latinos bear consistently disproportionate burdens of HIV/AIDS infection in the United States and Canada (Carballo-Diéguez, *et al.* 2005; Adam *et al.*, 2011), while rates of infection among gay/MSM Asians have at times exceeded those of white men (Choi *et al.*, 1996).

The high-risk profile of foreign-born gay men/MSM is connected to many different causal factors. Some factors may be related to the ostensible attributes of the immigrant's cultural group, such as hyper-masculinity (Carballo-Diéguez *et al.*, 2005), perceived inferiority, and gender nonconformity (Choi *et al.*, 1996; Somerville *et al.*, 2006) and expected or actual family rejection (Vega *et al.*, 2010). Other studies attribute risk to elements of the host society, such as systemic racism, the dominance of the English language, and the lack of integration into "mainstream" gay communities among ethnic minority gay men (Diaz *et al.*, 2001; Carballo-Diéguez *et al.*, 2005; Somerville *et al.*, 2006). Risk for HIV infection within this group, then, is often conceived in terms of double discrimination or multiple oppressions and vulnerabilities (Diaz *et al.*, 2001; Wilson and Yoshikawa, 2004; Vega *et al.*, 2010).

Fewer studies consider the logistical and emotional challenges of the migration process itself. Depending on each man's country of origin, education, and socioeconomic status, migrating might also involve de-skilling, precarious employment, and the need to support extended families (Somerville *et al.*, 2006; Adam *et al.*, 2011). Due to his immigration status, he might also have difficulty accessing health care (Adam *et al.*, 2011). Gay men/MSM who have moved from other countries might also be seeking to come out, find companionship and emotional connections, explore sexually, or gain entrance to local gay communities for the first time (Carrillo, 2004; Bianchi *et al.*, 2007; Vega *et al.*, 2010). At the same time, they may have had limited access to information about gay identities or HIV/AIDS in their home countries, and they may face exclusion or uneven power relations within mainstream gay communities (Han, 2007, 2008). Collectively, these factors can produce both structural circumstances and "states of mind" in which low self-esteem, depression, isolation, and hopelessness coalesce to lead to HIV risk factors such as unprotected intercourse and corollary behaviors such as substance abuse (Choi *et al.*, 1996; Carballo-Dieguez, 2005; Vega *et al.*, 2010). While studies on transnational migrations of gay men/MSM within the developing world are rare, studies on heterosexual migrant laborers in Africa and other regions (e.g. Kalipeni *et al.*, 2004) suggest that similar social factors (e.g. social upheaval) also apply to HIV risk behaviors in these locales.

Despite the emerging emphasis on the role of migration itself in HIV risk within the review literature, few of the studies included assess the effect of an intervention on immigrants *specifically*. Most choose to focus on a broad ethnocultural group (e.g. Latinos, Asians) that happens to include mostly foreign-born men. This theoretical orientation likely stems from the need for once-lacking cultural competency in HIV prevention interventions and the perceived need to address "harmful" characteristics of non-Euro-American cultures. At the same time, it may also elide important structural factors (e.g. stress and upheaval, isolation, and disempowerment) experienced by most immigrants to some degree regardless of origin. This chapter therefore seeks to not only assess the efficacy of the various intervention types used in immigrant populations, but to suggest how future interventions might be targeted and adjusted more specifically to immigrants and the migration process.

Intervention design

The six intervention studies reviewed here are drawn from a systematic search of the PubMed database using the terms "HIV," "intervention," "gay," or "MSM" and "immigrant," with specific ethno-racial group names (e.g. "Latino," "Asian") substituted for "immigrant." Criteria for inclusion were adoption of an HIV prevention intervention such as counseling or information provision, evaluation of the intervention using an experimental/control or pre-intervention/post-intervention assessment, location within the United States or Canada, and a majority of foreign-born men in the intervention group. In total, six studies were deemed acceptable for in-depth review. While several additional studies addressed planned or completed HIV prevention interventions geared toward gay/MSM immigrants (e.g. Conner *et al.*, 2005; Fernández *et al.*, 2009; Rhodes *et al.*, 2013; Martinez *et al.*, 2014), they were excluded due to the lack of an evaluative component. Among the six studies, five focused on Latino immigrants and one (Choi *et al.*, 1996) on Asian immigrants. Five were located in the United States with one in Canada. Five were located in large metropolitan areas (San Francisco, New York, Toronto) with large urban gay communities and robust histories of AIDS research and activism, while one (Melendez *et al.*, 2013) was located in two towns along the United States–Mexico border. Finally, only the Asian study had been conducted at the height of the AIDS crisis, while the Latino studies were relatively recent interventions. While three studies (Choi *et al.*, 1996; Carballo-Diéguez *et al.*, 2005; Melendez *et al.*, 2013) adopted control and intervention groups, the other three assessed a single group before and after the intervention (Somerville *et al.*, 2006; Vega *et al.*, 2010; Adam *et al.*, 2011). Two also performed post-intervention assessments at multiple temporal intervals to gauge the durability of the intervention (Carballo-Diéguez *et al.*, 2005; Vega *et al.*, 2010).

Despite some broad similarities in the siting and design of the interventions, they differed considerably in terms of depth, scope, and emphasis. Two interventions were relatively limited in scope. The first (Choi *et al.*, 1996) comprised a three-hour, single-session counseling intervention with recruited participants. Another (Somerville *et al.*, 2006) used facilitators trained at a two-day event to engage in short, informal information-sharing exchanges with community members. The other four interventions (Carballo-Diéguez *et al.*, 2005; Vega *et al.*, 2010; Adam *et al.*, 2011; Melendez *et al.*, 2013) were more extensive, involving 5–12 sessions over the course of several weeks. Almost all of the interventions (five of six) sought to promote safer sex information and skills, and most (four of six) sought to improve the participants' comfort with their sexuality or sexual identity. Two interventions (Vega *et al.*, 2010; Melendez *et al.*, 2013) engaged significantly with family issues and relationships, as well as societal homophobia. Another two interventions (Choi *et al.*, 1996; Vega *et al.*, 2010) also had explicit goals to develop social support within their respective ethno-cultural community. While the two earliest inventions (Choi *et al.*, 1996; Carballo-Diéguez *et al.*, 2005) were primarily psychological in nature (e.g. using professional counselors to increase self-efficacy through empowerment, better HIV knowledge, and an improved "state of mind"), the later interventions were based more on collective principles such as building family relationships or enhancing knowledge at the community level.

Culture as risk, clash, and shift

Many of the interventions included here emphasized the idea of culture as a risk factor, or as something that could be shifted or altered in order to promote better sexual health. While some of the interventions attempted to draw immigrant men away from elements of their culture of origin imagined to be destructive or harmful, others aimed to alleviate fallout from the imagined cultural clash between "traditional" Latino or Asian values and the unfamiliar norms (or discriminatory attitudes) of the gay and mainstream "host" communities. Many of the interventions discussed here positioned their subjects' cultures of origin as risk-promoting. In other words, the clash between conservative cultural values in ethnic communities and non-normative sexuality renders social and familial networks exclusive or even hostile. This clash may result in lowered self-worth among gay/ MSM immigrant men and therefore lead to increased risk-taking behaviors. The clash might also result in outright social rejection. Choi *et al.* (1996: 82) observe, for example, that "in most Asian cultures where male children have obligations to marry and perpetuate the family and its name, homosexuality is socially devi-ant and brings family dishonour." Similarly, many of the Latino studies explained that the dominance of culturally coded masculinity (i.e. a gender-normative *machismo* including gait, speech, mannerisms, and sexual practices) resulted in distance from family, lack of self-acceptance, and lack of reference points for understanding one's own sexual identity (Somerville *et al.*, 2006; Vega *et al.*, 2010; Melendez *et al.*, 2013). These negative cultural traits were often positioned in opposition to those of the mainstream gay community. While the white gay mainstream has established a "cultural niche" in the United States, ethnic minor-ity men might not have the same acceptance within popular culture due to their lack of spending power or cultural cachet (Somerville *et al.*, 2006).

At the same time, immigrant men may be trying to integrate into a new community – whether it is defined through sexuality, geography, or ethnicity – and could be trying to understand new community norms and how they differ from those learned in home and school settings within other countries (Vega *et al.*, 2010). Consequently, the mental health vulnerabilities that lead to risk-taking behaviors, such as isolation, depression, low self-esteem, downplaying risk, and failing to negotiate safer sex, are framed primarily as the product of a lack acceptance by self and family, and secondarily as the product of discrimination within mainstream society, or lack of acceptance into gay communities specifically (Choi *et al.*, 1996; Vega *et al.*, 2010; Melendez *et al.*, 2013).

Many of the interventions, then, are unsurprisingly aimed at shifting or bridging the imagined gaps between ethnic minority cultures and mainstream gay culture. At the same time, they often integrate aspects of immigrants' cultures to make the intervention translatable and to foster a sense of pride in one's sexuality and ethnic identity. Most of the interventions employ facilitators or leaders of the same ethno-cultural background due to acknowledged cultural differences. One study noted, for example, that "[c]ommunication in the Latino culture differs from that in other cultures. Latino communication is not linear as in the mainstream U.S. culture and care must be taken not to offend the Latino listener" (Somerville *et al.* 2006: 140). They may also employ culturally specific linguistic tropes such as *dichos* and *refranes* – sayings and mottos – to facilitate understanding of the intended messages (Carballo-Diéguez *et al.*, 2005). Yet another intervention had participants create *telenovela*-inspired comic strips to depict relevant sexual health issues for Latino MSM in Canada (Adam *et al.*, 2011).

However, the focus on culture in many interventions means that they seek to shift individual, internal states of mind, levels of HIV knowledge, and perceptions of identity. One intervention, for example, involved "increasing positive ethnic and sexual identity and to help acknowledge HIV risk behavior by discussing negative experiences of being both Asian or Pacific Islander and Homosexual" (Choi *et al.*, 1996: 85). Another sought a "reduction in internalized homophobia, self-acceptance, self-esteem and self-comfort with sexual identity," based on the notion that a confirmed identity and intent to disclose it to family reduces risk (Melendez *et al.*, 2013: S493). While achieving such goals might improve the individual psychological outcomes that act as intervening factors in HIV risk, they do not necessarily account for the more systematic circumstances and insecurities brought about by the process of immigration itself.

Addressing the immigration process

To a lesser extent, the selected interventions also addressed the risk contexts and situations arising from the migration process more generally. Three of the interventions (Somerville *et al.*, 2006; Vega *et al.*, 2010; Adam *et al.*, 2011) were tested specifically among immigrants (rather than second- or third-generation participants) and one (Adam *et al.*, 2011) was specific to recent immigrants who had arrived in the past three years. Even so, only one (Adam *et al.*, 2011) specifically emphasized skills for coping with the challenges of immigration and resettlement. Studies in both the international (Bianchi *et al.*, 2007) and intra-national (Egan *et al.*, 2011; Lewis, 2014) contexts indicate that gay men and MSM frequently experience a sense of freedom after moving, but also social upheaval and insecurity. While sexual exploration for immigrant men may therefore be highest shortly after moving, social supports may be less developed, and knowledge of how to negotiate sexual interactions is potentially limited.

To some extent, the intervention studies support this theory. Some argue that gay men/MSM might migrate specifically to establish a gay identity, which could be seen as a source of resilience and self-actualization rather than risk (Carballo-Diéguez, 2005; Melendez *et al.*, 2013). At the same time, gay men and MSM who have moved for economic reasons (and may not be out or looking to come out) still negotiate the challenges of negotiating sexuality (and sex itself) in unfamiliar settings. These challenges include disrupted social networks, isolation, and anxiety associated with migration (Adam *et al.*, 2011), as well as the need for companionship and putting oneself at risk for a moment of acceptance (Somerville *et al.*, 2006). In addition, structural issues such as lack of education, language skills, and precarious employment can also affect exposure to HIV among immigrant MSM. Although these factors are often framed as "complicat[ing] attempts at implementing more generic prevention programs" (Vega *et al.*, 2010: 4), they should also be seen as risk factors in and of themselves. One study cites a mean yearly income of just above $20,000 dollars among participants, a median income of $12,000, and only three-fifths of participants working for pay (Carballo-Diéguez *et al.*, 2005). When coupled with low education levels and lack of familiarity with their surroundings, the poverty experienced by immigrant men can also result in exploitative relationships (Somerville *et al.*, 2006). In addition, perceived immigration status (e.g. illegality or refugee status) may compromise access to health services, employment, educational opportunities, and social networks (Adam *et al.*, 2011).

A few of the interventions emphasized overcoming the challenges of migration as opposed to those associated with culture. Indicating that prevention interventions need to be anchored within the larger migration context rather than culture, the researchers for the *La Familia* intervention "were eager to have an intervention that highlighted each member's life story and nationality while impacting and integrating into a larger network" (Vega *et al.*, 2010: 4). The *Mano en Mano* study (Adam *et al.*, 2011) specifically combined post-immigration life skills with HIV prevention skills, seeking to facilitate community involvement and activism among its participants. Rather than executing a psychological intervention aimed at changing risk indicators (e.g. self-esteem) or perceived cultural attributes (e.g. internalized homophobia), the sessions for *Mano en Mano* covered practical topics ranging from local gay cultures and communities, dating, cybersex, and social and health services for newcomers. Sexual activity, meanwhile, was positioned as something that could increase following immigration but whose safety could be compromised by the stressors and challenges that accompany the resettlement process. Similarly, the *SOMOS* intervention (Somerville *et al.*, 2006) teaches participants to be aware of the social inequalities surrounding immigration and to avoid compromising safety in order to gain companionship or perceived acceptance.

Conversely, the Choi *et al.* (1996) intervention did not address migration challenges and found that the group counseling sessions (focused on empowerment amid adverse aspects of Asian culture) had a statistically significant effect only for Chinese and Filipino participants, who represented the most established Asian groups in San Francisco. In contrast, they found that "other homosexual [Asian and Pacific Islander] groups may not have a solid peer network and consequently have less opportunity to socialize with homosexual men of a similar background" (Choi *et al.*, 1996: 85). The Carballo-Diéguez *et al.* (2005) intervention, which had a similar emphasis on counseling-based self-empowerment strategies rather than developing post-migration skills and resiliencies, saw no statistically significant intervention effect.

Intervention evaluation

Most of the interventions indicated some degree of success. Typically, the success was measured in terms of a reduction in a risk behavior (e.g. unprotected anal intercourse) or an improvement in resiliencies such as intent to come out, more social supports, or self-esteem. Most of the interventions also required the self-report of a "risk" – usually unprotected sex – at a baseline assessment. This baseline allowed the researchers to evaluate whether the intervention successfully reduced HIV risk. As is typical in health intervention research within difficult-to-reach populations, most of the studies conveyed tensions, challenges, and compromises in both sampling and evaluation methodologies. All of the studies employed convenience-based samples due to the specificity of the populations being researched, or funding requirements that demanded particular demographic characteristics. Since this technique recruits participants through specific venues (e.g. Vega *et al.*, 2010), social networks (e.g. Somerville *et al.*, 2006), or health service providers (e.g. Choi *et al.*, 1996), there is strong potential for the self-selection of participants who are already involved in community networks, more receptive to health information, and potentially more knowledgeable about HIV prevention before beginning the intervention. In a venue-based sample (Vega *et al.*, 2010), for example, 20 percent of the Latino immigrant men recruited had completed university, which is not representative of the overall lower educational attainment among Latino immigrants. The self-selection of established community members could lead to greater uptake of the intervention among certain groups (as in Choi *et al.*, 1996), but could also potentially dull the measurable effect of the intervention (as in Carballo-Diéguez *et al.*, 2005).

Each study either compared a group that received the intervention versus one that did not (e.g. Choi *et al.*, 1996; Carballo-Diéguez *et al.*, 2005) or measured various risk indicators within a single group before and after the intervention (Somerville *et al.*, 2006; Vega *et al.*, 2010; Adam *et al.*, 2011; Melendez *et al.*, 2013). Most methodological intervention literature suggests, however, that comprehensive evaluation requires a quasi-experimental design employing pre- and post-intervention data for both a case (intervention) group *and* a control group. Different from the other studies, the *Promotores* intervention (Somerville *et al.*, 2006) used a popular opinion leader (POL) mode of health promotion to diffuse health messaging into the Latino immigrant MSM communities in Vista

(CA) and McAllen (TX). Its evaluation involved surveying one set of available respondents before the initiation of the intervention, and then a different set of respondents after the intervention. The evaluation, however, treated the two groups as equivalent community cross-sections despite not reaching the 15 percent population benchmark recommended for POL interventions. The intervention had also relied on the same *Promotores* who delivered the health information to perform the evaluative surveys, introducing potential bias. The Melendez *et al.* (2013) study had a similarly flawed evaluation process that assigned participants to the intervention group, even if they had received only a few sessions of the intervention rather than completed the full intervention.

Interestingly, the studies that claimed the highest level of rigor based on their case-control designs fared worst in terms of quantifiable success (see Table 5.1). Neither the Choi *et al.* (1996) study of brief group counseling for Asian immigrant MSM nor the Carballo-Dieguez *et al.* (2005) study of a longer-term empowerment-based intervention geared toward HIV risk reduction for Latino MSM found a statistically significant difference in the reduction of unprotected anal intercourse or other measures of self-reported sexual risk-taking between the control and intervention groups. The Asian study did, however, observe a significant difference in the reduction of sexual partners, as well as significant increases in HIV-related knowledge and skills. While the Asian study attributes the limited measurable success to the brevity of the intervention, the Latino study cites possible selection bias of already-empowered men or the persistent awareness of being studied within both the intervention and control groups, leading to similar behavioral changes in both groups (i.e. a Hawthorne effect).

The pre- versus post-intervention studies were more diverse in terms of both the variables measured and levels of success reported. Half the studies measured proximal variables such as unprotected anal intercourse (i.e. the "gold standard" of HIV intervention success), and most of these added intervening variables such as loneliness and self-esteem (Somerville *et al.*, 2006; Adam *et al.*, 2011; Melendez *et al.*, 2013). A fourth intervention evaluation measured only distal variables such as HIV knowledge and intended behavior (Vega *et al.*, 2010). Adam *et al.* (2011) reported a significant reduction in unprotected sex from the baseline, but no reduction in loneliness (the sole intervening variable measured). Somerville *et al.*'s (2006) POL study found a reduction in unprotected sex in the second year of the study only, and improvement in one measure of basic HIV knowledge (non-transmission through clothes or hats) in the second year only (results were measured year by year). While Melendez *et al.* (2013) observed statistically significant reductions in measures of sexual intent (e.g. using sex to feel better) and other more distal variables (e.g. ashamedness of being gay), they did not observe significant increases in condom use – a more proximal measure of risk. Vega *et al.* (2010) saw significant reductions in a number of negative intervening variables (e.g. numbers of partners, types of partners) and improvements in measures of HIV knowledge and coping ability/community membership) despite observing low levels of these at the baseline pre-intervention assessment.

Table 5.1 Studies evaluating HIV prevention interventions aimed at immigrant men who are gay/MSM

Study (intervention name)	Location	Year(s)	Ethnic group	n	% immigrant	Intervention type	Intervention effects*
Choi et al. (1996)	San Francisco, CA, USA	1992–4	Asian and Pacific Islander	329	67	3-hour group counseling session	*Control vs. intervention group at 3 mos.:* 6.4 vs. 3.9 mean no. of sexual partners 7.53 vs. 7.41 F-stat. for AIDS knowledge 2.26 vs. 2.43 F-stat. for AIDS anxiety
Carballo-Diéguez et al. (2005): *Latinos Empowering Ourselves*	New York City, NY, USA	1998–2001	Latino	180	75	8 2-hour group activity sessions	*Control vs. intervention group at baseline:* 100% vs. 100% reporting UAI *Control vs. intervention group at 2 mos.:* 46% vs. 54% reporting UAI *Control vs. intervention group at 8 mos.:* 67% vs. 70% reporting no UAI *Control vs. intervention group at 14 mos.:* 67% vs. 70% reporting no UAI (no statistically significant differences)
Somerville et al. (2006): *Promotores*	McAllen, TX, and Vista, CA, USA	2003–5	Latino** (mostly Mexican)	766	100	2-day training program for 37 Youth Latino Promotores (YLPs)	Intervention group at baseline vs. post-intervention (rolling assessment): 34% vs. 17% agree sharing clothes or hats transmits HIV*** 34% vs. 50% using condoms as receptive partner*** 58% vs. 74% report giving and receiving oral sex without condoms***
Vega et al. (2010): *SOMOS*	New York City, NY, USA	2002–6	Latino (Central and South American, Puerto Rican)	113	NA	Series of 5 group counseling sessions	*Intervention group at baseline vs. 3 mos.:* 35.98 vs. 40.08 score for AIDS knowledge 1.62 vs. 1.18 mean no. of sexual partners 2.17 vs. 1.97 mean no. of partner types 5.33 vs. 4.35 HIV risk score 15.18 vs. 16.17 self-esteem score

(Continued)

Table 5.1 Studies evaluating HIV prevention interventions aimed at immigrant men who are gay/MSM (Continued)

Study (intervention name)	Location	Year(s)	Ethnic group	n	% immigrant	Intervention type	Intervention effects*
Adam et al. (2011): Mano en Mano	Toronto, ON, Canada	2009–10	Latino (Mexican, Venezuelan, Colombian, Peruvian, Ecuadorian)	40	100	1-day group information session + 4 2-hour group information sessions	11.44 vs. 12.19 reported social supports *Intervention group at baseline vs. 6 mos.:* 1.62 vs. 1.07 mean no. of sexual partners 11.44 vs. 12.19 reported social supports *Intervention group at baseline vs. 2 wks.:* 20% vs. 8% reporting UAI w/ reg. partner 13% vs. 3% reporting UAI w/ cas. partner
Melendez et al. (2013): La Familia	San Francisco, CA, USA	2010–11	Latino (Mexican, Central and South American)	49	100	6-week, 12-session program	*Intervention group at baseline vs. post-intervention (end of last session):* 31% agree w/ being ashamed of being gay vs. same 31% disagree w/ being ashamed *Intervention group at baseline vs. 2 mos.:* 46% report more friends who understand problems 11% feel part of gay community in Mexico vs. same 11% not feeling part of gay community in Mexico *Control vs. intervention at 2 mos.:* No effect vs. fewer using sex to feel better 27% vs. 58% felt more comfortable w/ sexuality 30% vs. 55% had tools to come out to family 19% vs. 49% had come out to a family member

* Only differences statistically significant at 0.05 or less are shown unless otherwise noted.
** 18–25 year-olds only.
*** Year 2 (2004–5) results only (not significant for Year 1 of intervention).

Discussion

A straightforward assessment of these six interventions is not an easy proposition. Each employs a different conception of HIV risk and different combinations of proximal and intervening variables. In addition, many of the studies have significant methodological limitations. Each, however, can generally be evaluated in terms of degree of difference achieved (between control and intervention group) or change observed (pre- vs. post-intervention), meaningfulness of the indicators for which differences and changes are observed, and reliability of the intervention study in terms of both methodological design and the emphasis on culture vs. migration.

Vega *et al.*'s (2010) *SOMOS* study fared the best in terms of producing statistically significant changes across a wide variety of indicators, but none of those indicators measured actual behavioral change. In contrast, Adam *et al.*'s (2011) *Mano en Mano* study showed the clearest reduction in unprotected anal intercourse, though it was based on a small sample (n = 40) and did not feature a control group. Notably, the *Mano en Mano* intervention was the one most focused on post-immigration processes and skills (e.g. finding health services, housing and settlement challenges, dating, and acclimating to Canada, Toronto, and the Toronto gay community) rather than the culture, identity, or knowledge base of the participants. The two intervention-control studies that attempt to shift these latter attributes through psycho-social interventions are also the ones that recorded the fewest measurable effects. Meanwhile, the remaining studies – all of which employ more collective approaches – showed moderate success in terms of promoting community networks and positive identity formation among immigrant MSM, but had methodological limitations that might have limited actual behavioral change (or measurement of it). These drawbacks provides preliminary evidence of the need for sustained, long-term interventions among immigrant MSM that are designed as randomized controlled trials or that at least take additional precautions to avoid sampling bias.

There also seems to be evidence of a need for greater attention to the migration experience in HIV interventions geared toward ethnic minority men. First, the most quantitatively successful interventions were those that used a 100 percent foreign-born sample, eliminating the effect of second- and third-generation participants who may be less affected by the intervention. These more established groups might already be quite adapted to North American cities, their gay communities, and ways of negotiating various ethnic and sexually based social networks, regardless of any "negative" or harmful traits associated with their ethnic group's culture. Second, many of the interventions emphasized cultural constructs (e.g. intent to come out, family support for coming out) that might not necessarily resonate with ethnic minority and immigrant MSM. Not only might these men not ascribe to the notion of coming out (see Jivraj and de Jong, 2011), but doing so would not necessarily inspire the confidence or resiliency to lead them to safer sex behaviors. Third, recent studies – including Somerville *et al.* (2006) in this review – suggest that HIV risk among immigrants tends to be associated with depression, isolation, and stress associated with the process of migration and resettlement itself, not just the cultural characteristics of an immigrant group (Carrillo *et al.*, 2004; Bianchi *et al.*, 2007; Lewis, 2014). As such, logistical and practical skills of adaptation and integration focused on housing, healthcare, and dating may be as important as gauging psycho-social attributes such as internalized homophobia, family support, or knowledge of HIV/AIDS. Interestingly, the intervention most focused on these aspects (Adam *et al.* 2011) also showed the clearest reduction in sexual risk behavior. Continued emphasis on community-designed, community-based, and rigorously tested interventions addressing post-migration challenges will therefore be central to the future analysis of HIV risk in the growing immigrant MSM population.

Geographers in particular can make contributions to the understanding of how interventions geared to sexual minority groups within immigrant populations are developed, implemented, and evaluated. Population geographers, for example, have begun to look at how vulnerabilities can emerge and be reinforced *across* the life course (Heikkilä, 2005; Bailey, 2009; Lewis, 2014). HIV prevention intervention for immigrants, then, need not only address the risks of a particular place (e.g. poverty and poor housing in immigrant neighborhoods) or particular populations (e.g. internalized homophobia in Latino men), but also past traumas that often prompt migration and the social upheaval and loss of supports that accompany it. Geographers have paid considerable attention to how specific places and place experiences, including war zones and urban ghettoes, are internalized in ways that can diminish mental health, well-being, and health-seeking behavior (Dear and Wolch, 1987; Davidson and Milligan, 2004). Closer analysis of these geographies can help developers of interventions better understand the sequences of landscapes that gay immigrants encounter across the life course, as well as the finer distinctions associated with various origins, destinations, and trajectories of migration. Relatively established groups of established Cuban immigrants living in Miami, Florida, may have far different intervention needs (e.g. club drug use prevention) than those of recently arrived Mexican men in North Carolina (e.g. social isolation and poverty).

HIV prevention interventions geared toward gay immigrant and ethnic minority men might also benefit from more geographically influenced analyses of where interventions are implemented. Amid the growing suburbanization of the immigrant population and the tendency for newly arrived men to remain with families or social contacts in ethnically concentrated communities outside of the city centre, intervention leaders can no longer look to centrally located gay communities or ethnic communities as the sole point of intervention for these populations (Zablotska *et al.*, 2011). In addition, the geographies of gay communities themselves are changing, with affinities for the traditional "gay village" decreasing in many cities (Rosser *et al.*, 2008; Nash, 2013; Brown, 2014). Geographers might therefore be helpful in not only addressing where the relevant populations are located and where interventions can be accessed, but also in determining whether traditional single-point interventions (e.g. group counseling at a local clinic) are appropriate, or if a more diffuse approach (e.g. a popular opinion leader intervention) is needed. Finally, geographers might also assist with evaluating HIV prevention interventions. Geo-coded analyses could allow health professionals to see the geographic reach of diffusion-based interventions or assess whether the efficacy of single-point interventions varied based on location of residence (e.g. near gay nightlife venues, in wealthy versus deprived areas) for the intervention participants. Integration of these techniques has the potential to lend geographic contingency and complexity to the current focus on culture in sexual health interventions for sexual minority immigrant men.

6 Making a place for health in vulnerability analysis

A case study on dengue in Malaysia and Brazil

Sarah Dickin

Sustainable development has become an urgent global priority. The outcomes of rapid economic growth and population increase, characterized by intensive energy use, food production, urbanization, and waste generation, are gradually being observed. These human pressures are disrupting the life-supporting properties of the "Earth System" which our societies rely on (Rockström *et al.*, 2009). Climate change, depletion of freshwater resources, and degradation of aquatic and terrestrial ecosystems are all evidence of widespread shifts in Earth System processes, with some researchers calling for a new geological era termed the "Anthropocene" (Crutzen, 2002).

These unprecedented global environmental changes are interlinked with social, economic, and political crises and inequalities, and together pose a growing threat to the health of the world's population (ISSC and UNESCO, 2013). A range of impacts on health have been identified as occurring through direct and more indirect mechanisms. Direct impacts include changes to disease transmission patterns, such as shifts in the seasonality and range of vectors and the emergence of new infectious agents such as *Cryptosporidium* and Nipah virus. Indirect pathways often entail social determinants of health, such as displacement of populations leading to mental health impacts and loss of livelihoods (Kovats and Butler, 2012). Overall, these changes are systemic in nature, operate across large spatial and temporal scales, and are difficult to remediate, highlighting a transition towards a new category of intractable health challenges (McMichael, 2014).

Global environmental change presents obstacles for population health research, as the impacts differ from more classical environmental health hazards in several ways. Exposures are not discrete or easily monitored, compared with toxins such as particulate matter inhaled due to air pollution (Figure 6.1). In addition, these changes impact whole communities or populations, requiring study at an ecological rather than individual level (Krieger, 2014). Many of these changes, such as the destruction of ecosystems providing regulating and provisioning services, may result in new states or conditions that are very difficult to remediate, such as oil contamination of the Ogoniland wetlands in the Niger Delta (UNEP, 2011). In addition, these changes are disproportionately impacting those populations contributing little to creating the problems and with the least to gain.

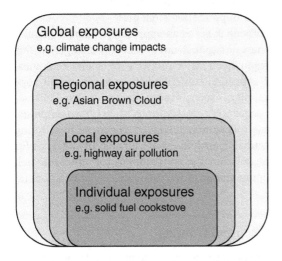

Figure 6.1 Different scales of exposures: individual and local exposures are more direct-acting (McMichael *et al.*, 2003).

The health implications of global environmental change are further character-ized by complex interactions and dynamics (Gatrell, 2005). Most pathways do not operate independently, with cumulative effects and interdependencies that make identifying cause and effect challenging (Costello *et al.*, 2009). Processes operate across multiple spatial scales, ranging from local hygiene practices to watershed hydrology in the case of diarrheal disease (Eisenberg *et al.*, 2007; Myers and Patz, 2009). Across temporal scales system changes may occur gradu-ally over extended periods, or in sudden shifts, such as the rapid spread of Chikungunya, an emerging mosquito-borne disease, throughout the Caribbean (Morens and Fauci, 2014). This complexity is also manifest as non-linear behav-iors that are difficult to anticipate and have unintended consequences. The result may be "surprising" disease outbreaks or the emergence of new infectious diseases, illustrated by the expansion of schistosomiasis with the construction of dams and irrigation projects, or increased gastrointestinal illness associated with road construction (Eisenberg *et al.*, 2006).

In the face of these challenges, new approaches and tools are needed to address emerging threats to population health and to develop appropriate interventions. However, health researchers and practitioners have been slow to engage with the significance of these global transformations and with the greater sustainable development agenda. Many approaches address only a subset of the considera-tions needed, due in part to a traditional focus on individual-level determinants (Batterman *et al.*, 2009). Social and environmental determinants of health must be considered together to manage these emerging threats to health as previously advocated in the Millennium Ecosystem Assessment (Parkes and Horwitz, 2009; Plummer *et al.*, 2012). In addition, many of the contributing factors occur outside of the health sector, which requires inter-sectoral collaboration across decision-making levels to find sustainable health promotion strategies.

This chapter presents vulnerability analysis as an approach to address population health challenges in the context of global environmental change. In this discussion vulnerability is conceptualized as a property of social-ecological systems, comprising elements of exposure, susceptibility, and capacity to cope with a health hazard. Thus vulnerability is seen as multi-dimensional and characterized by complex relationships occurring across various scales. Two empirical examples are presented to illustrate this approach applied to dengue in differing geographical contexts and the potential contributions of vulnerability analysis for population health intervention research. Finally, opportunities for future research are presented emphasizing the coping and adaptive capacities of human populations in the face of global change.

Vulnerability and health: conceptual framing

The concept of vulnerability has been extensively applied to investigate the response of social-ecological systems to global change. While there remain contested understandings, generally vulnerability is defined by a system's exposure and susceptibility to stress or shocks, and its capacity to cope or adapt in these conditions (Adger, 2006). Current research on vulnerability is influenced by two disciplinary theories, emphasizing a hazard-centric or people-centric framing (Eakin and Luers, 2006). Advances that conceptualize vulnerability in a broader sense, as a property of a coupled human-environment system, have built on elements from these research traditions originating from the literature on natural hazards and disasters and development studies. Integrated frameworks have contributed to the current understanding of vulnerability to global environmental change and here the concept of vulnerability is further developed for applications in population health.

Vulnerability analysis emerging from work on natural hazards is known as a "risk-hazard" approach and has been extensively used in the technical literature on disasters such as flood management. The risk-hazard approach takes the view that almost all types of natural hazards have differential outcomes, most impacting marginalized groups based on where they live and their resources (Burton *et al.*, 1978). This perspective focuses on a hazard as a starting point and is commonly applied to assess the potential for loss or harm from a range of biophysical threats, such as climate change impacts. Using this concept risk is seen as a function of a hazard and the vulnerability of human populations to its impacts, where a hazard is considered a potential event, phenomenon or activity with negative consequences (UN ISDR, 2004). This approach emphasizes identification of the types of hazards, their consequences, and when and where they are likely to occur (Eakin and Luers, 2006). The framing is consistent with the IPCC definition of vulnerability as a function of exposure, sensitivity, and adaptive capacity (McCarthy *et al.*, 2001). However, this approach has been criticized for neglecting human systems in determining the outcome of a hazard, such as political and economic dimensions (Hewitt, 1983).

An alternative approach places the focus on people, and seeks to determine differences in the vulnerability of individuals, groups or communities and the reasons for this variation. Political economy frameworks are applied to examine social, economic, and political relations that create differential vulnerability (e.g. Watts and Bohle, 1993). In particular, this framing has been widely used in development literature to characterize multi-dimensional aspects of poverty and livelihoods (DFID, 1999). Development of a political economy approach to vulnerability was influenced by Sen's work to explain food insecurity during famines as a lack of entitlements (1981). This view describes people's entitlements to assets and resources as an essential element in their vulnerability to a hazard, rather than the lack of food such as in the case of a famine. Applying a political economy approach has made contributions to highlighting how social inequities shape vulnerability, including access to resources and institutional structures and processes (Adger, 2006). Emphasizing coping capacity as part of this approach has improved understanding of variable capacities to respond and recover from stress or a hazard (Adger and Kelly, 1999). Contributions of the physical environment are not addressed but political ecology frameworks have responded to this critique by providing a greater representation of these elements. However, these must be linked to differential vulnerability outcomes to avoid general descriptions of inequalities (Eakin and Luers, 2006; McLaughlin and Dietz, 2008).

Researchers interested in the implications of global environmental change have attempted to bring these traditions together to represent both social and physical dimensions of vulnerability. Advances using social-ecological systems-based approaches, which emphasize inextricable links between humans and their environment, have been the basis of several conceptual frameworks integrating elements of these disciplinary traditions. A framework proposed by Turner *et al.* (2003) develops a coupled human-environmental description of vulnerability that is place and time specific, while making explicit links to other scales. Birkmann *et al.* (2013) build on several integrated approaches to describe a "MOVE" framework to assess multi-faceted vulnerability, incorporating aspects of complex systems such as feedback loops. These integrative framings of vulnerability have been used to study a range of hazards, including water shortages due to urbanization, loss of livelihoods from ecosystem disturbances, and natural disasters linked to climate change. It is important to note that some researchers argue these approaches cannot be truly brought together due to epistemological differences (O'Brien *et al.*, 2007), however applications of socio-ecological framings of vulnerability are popular due to their ability to show disparate types of information.

Applications in population health

Despite the relevance of global environmental change for health and well-being, vulnerability frameworks have not been widely applied to this context. One example where vulnerability analysis has been applied is research exploring the health impacts of climate change (Ebi *et al.*, 2006; Confalonieri *et al.*, 2013). However, the health implications of global environmental change extend beyond climate change, and examining interactions between a range of social and environmental changes is critical to providing a holistic understanding on which to base interventions. There are several parallels linking vulnerability concepts to population health that indicate the potential to generate new insights. Drawing on these ideas, this chapter seeks to apply vulnerability analysis to the case of water-associated disease.

Population health approaches consider a full range of determinants, including social, political, economic, cultural and ecological factors (Young, 2005). In addition, to the influence of diverse environments, population health emphasizes the theme of coping capacity (Lindsay, 2003). Vulnerability analysis provides an organizing framework to examine these determinants in the context of a changing planet, examining routes of human exposure to health hazards, and a population's susceptibility and ability to cope with their impacts (Few, 2007). While several approaches seek to integrate social and environmental perspectives to study population health (e.g. McMichael, 1999; Krieger, 2001b), a vulnerability framework explicitly incorporates ecosystem processes which are overlooked in some ecological perspectives (Parkes and Horwitz, 2008). This is required for understanding the role of global environmental change in influencing health, and is emphasized in the Ottawa Charter and the Millennium Ecosystem Assessment, which advocate that health promotion cannot be separated from other objectives, including ecosystem management and conservation (WHO, 1986; Corvalán *et al.*, 2005). In addition, population health approaches describe health outcomes that arise from the complex and dynamic interactions among determinants (Kindig and Stoddart, 2003). Similarly, analysis of vulnerability as a property of a social-ecological system allows descriptions of complexity that are central to understanding processes of global environmental change, such as cross-scale interactions and feedback loops. Placing health more explicitly among concepts of vulnerability illustrates how this perspective can be used to assess the people and places most vulnerable to health hazards while drawing attention to the uneven impacts of global environmental change (Marmot, 2007).

Another relevant component of vulnerability analysis is the emphasis on providing relevant information for decision-makers which is well aligned with the goals of population health intervention research. Vulnerability assessment is designed to raise questions regarding which populations can better manage health threats in a dynamic and changing world, in order to support policy-makers in better targeting of resources and interventions. In this respect a range of tools have been developed to engage and communicate with decision-makers, such as indices and mapping approaches (e.g. Plummer *et al.*, 2012). Using straightforward approaches, vulnerability assessment aims to communicate the complexity of health threats in a specific place or population posed by current and, importantly, future global environmental change. Providing stakeholders with a more holistic understanding can improve collaborative efforts towards shared sustainable development objectives of promoting healthy populations and ecosystems.

Assessing vulnerability

Vulnerability analysis is situated at the nexus of academic work and policy needs; however, measuring vulnerability is an area of ongoing debate. The challenge centers on empirical methods to represent complex social and physical interactions in ways that provide useful information for context-specific interventions. Although there is a rich range of frameworks describing dynamic processes that characterize vulnerability to global environmental change, these have been criticized for providing conceptual findings that are challenging to translate into relevant information for decision-makers (Luers, 2005). Some important dimensions of vulnerability are particularly intangible or difficult to capture, and thus some assessments cannot be related back to theoretical descriptions (Miller *et al.*, 2010). Despite these challenges there is significant interest in operationalizing vulnerability concepts to develop indicators and assessment tools. Well-defined indicators can be used as tools for comparative analysis, the identification of strategies to reduce vulnerability and increase coping capacity, and the better targeting of interventions. To empirically assess vulnerability a range of approaches for selecting and weighting indicators to create indices have been proposed (Hinkel, 2011; Hagenlocher *et al.*, 2013). Validation of indicators is an additional challenge due to the holistic nature of vulnerability; however, options including using an independent dataset as a proxy for vulnerability or internal validation of index construction have been applied (Fekete, 2009).

Traditionally, probabilistic risk factor analysis has been commonly used in population health research. However, a challenge to addressing the emerging category of health hazards linked to global environmental change is that assessments examining linear relationships do not adequately address the interacting pathways mediating health outcomes (Eisenberg *et al.*, 2007). In this case, developing analyses that estimate risk may play less of a role than elucidating key pathways that can be targeted (McMichael, 2014). For example, in the case of climate change in the Great Lakes Basin where there is high uncertainty surrounding future projections, vulnerability assessment was conducted to identify entry points for using climate information (Brown and Wilby, 2012). However, to engage stakeholders and provide useful information, vulnerability assessments must be conducted within a defined system, place or population, while remaining aware of linkages to other scales (Turner *et al.*, 2003). Moreover, narrowing the analysis to focus on specific health outcomes provides more concrete information for decision-making needs (Few, 2007).

Examples of vulnerability assessment applied to dengue

The application of vulnerability assessment to a health context is illustrated using the example of dengue in two case studies. The water-associated disease index, developed at the United Nations University Institute for Water, Environment and Health, was used to map vulnerability in two geographic contexts in Malaysia and north-eastern Brazil. In this approach, vulnerability was defined as the propensity to be adversely impacted by a water-associated health hazard, described by inter-actions between exposure, susceptibility, and resilience. These studies illustrate an integrative vulnerability assessment approach incorporating social and ecological processes and interactions that describe water-associated disease transmission. The first case study from Malaysia illustrates the use of the water-associated disease index to empirically assess vulnerability to dengue, while the second study builds on this approach incorporating additional system dynamics.

Dengue is a mosquito-borne disease that affects approximately half the world's population living tropical and sub-tropical regions where it is transmit-ted. An estimated 400 million infections occur annually, with most cases reported in Southeast Asia, the Americas and the western Pacific (Gubler, 2014). The disease causes flu-like symptoms which can develop into life-threatening complications. Because *Aedes* mosquitoes carrying the virus breed in water-filled containers in proximity to human populations, dengue is considered a water-associated disease. Dengue is linked to water through interactions between social, ecological, economic, and political domains, for example the use of water storage containers is prevalent in areas not adequately served by a public water supply, that experience unreliable water service, frequent water-scarce conditions, or that experience economic barriers to using a water supply. As there is no specific treatment for dengue, interventions focus on vector control carried out by households and communities (e.g. covering and cleaning water containers) and public health officials (e.g. insecticide fogging, educa-tional campaigns, surveillance).

As the most rapidly expanding vector-borne disease, dengue has grown more than 30 times in the last several decades due to large-scale changes, such as urbanization and global travel combined with ineffective vector control (Gubler, 2014). This expansion has highlighted a poor understanding of the complex social and ecological systems in which dengue is transmitted and provides an impetus for examining this health challenge in the context of global environmen-tal change. In addition, a recent WHO strategy recognizes a need to identify vulnerable groups and geographical areas to enhance effectiveness and sustaina-bility of dengue prevention (WHO, 2012a).

To apply the water-associated disease index, a conceptual framework describ-ing dengue transmission was used to identify indicators of exposure, susceptibil-ity, and resilience based on evidence in the literature. Exposure was defined as the conditions that support the presence and transmission of the disease agent, in this case *Aedes* mosquitoes that carry dengue. Susceptibility referred to sensitiv-ity when exposed to a hazard, which may include social, economic, and political conditions (Birkmann *et al.*, 2013). Resilience was defined as an ability to cope or adapt to a hazard, for example access to healthcare would allow a population to better respond to a dengue outbreak. As it is difficult to differentiate some resilience and susceptibility indicators in empirical assessments (e.g. education level), these indicators were combined in this analysis (de Sherbinin, 2014). In both case studies data were identified to populate these indicators using freely accessible information (Figure 6.2).

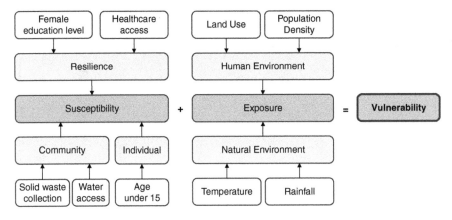

Figure 6.2 Indicator components included in the water-associated disease index applied in north-eastern Brazil

While the conceptual framework provides a comprehensive view of conditions contributing to disease transmission, data limitations constrain what can be included in a quantitative index. While certain relevant datasets were unavailable for inclusion in the index, such as housing quality, this limitation provides information to decision-makers on data needs to improve future work. In each case data were standardized to a range from 0 to 1 and translated into a consistent format in order to combine layers into an output that could be visualized using a geographic information system (GIS). While indicator components were equally weighted, a weighting methodology was applied that assigned a greater weighting to the resulting exposure indicator. Indicators of exposure and susceptibility were then used to build and visualize the vulnerability index. As vulnerability is multi-dimensional, validation of a vulnerability assessment is limited, but various types of evidence can be used as a proxy (Fekete, 2009; Srinivasan *et al.*, 2013). In these studies dengue rates were used as a proxy for vulnerability to validate the index outputs.

In Malaysia the water-associated disease index was applied to map vulnerability at a national level, providing outputs that illustrate regional trends relevant for decision-makers working on this scale (Dickin *et al.*, 2013). An important output of vulnerability assessment is identifying places with highest vulnerability in order to target interventions. The results of this analysis indicate that regions of high vulnerability are located in urban areas where exposure conditions, such as high population density, are very suitable for dengue vectors. Vulnerability assessment can also provide insight into interacting social and ecological processes that are less obvious to decision-makers. For example, eastern Malaysia is strongly affected by the monsoon season, which brings heavy rainfall to the coast, possibly washing away outdoor mosquito breeding sites, while in drier months, exposure in this region increases, due to more moderate rainfall and ideal temperature conditions (Figure 6.3).

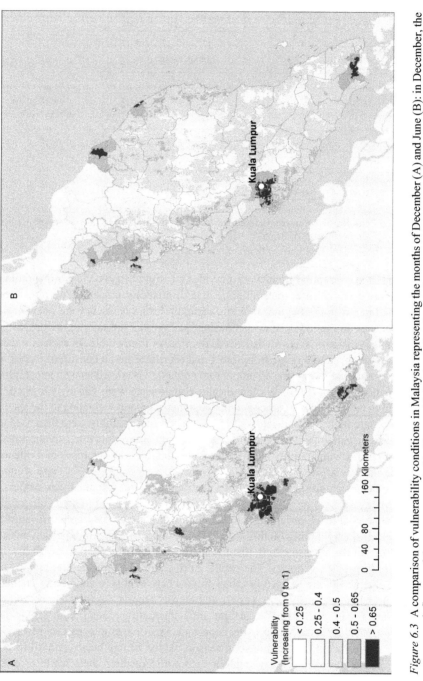

Figure 6.3 A comparison of vulnerability conditions in Malaysia representing the months of December (A) and June (B): in December, the influence of the monsoon period in Eastern Malaysia can be observed

These strong seasonal fluctuations in climate conditions interact with high levels of susceptibility and lower coping capacity in this region, compared with larger urban areas such as Kuala Lumpur. For example, lower levels of female education may reduce the ability to understand health promotion messages relating to water storage around the home. The seasonal peak in dengue cases often observed during the inter-monsoon period may be linked in part to this interaction which creates higher vulnerability conditions. In terms of public health planning, this pattern indicates that timing dengue interventions based on seasonal dynamics could be tested in the region. Moreover, monsoon patterns may become more sensitive to climate change in the future, an area for future research.

In contrast, some areas of Malaysia have consistently low vulnerability due to low exposure from unsuitable land uses for *Aedes* vectors, as well as low population density. No vulnerability is observed in areas where temperatures are too low, mainly the mountainous regions in central Malaysia. However, highly susceptible populations may become a concern if exposure conditions change, such as temperature increases linked to climate change. Overall, the vulnerability mapping results suggest that national vector control plans should be adjusted by region based on place-specific processes that increase vulnerability.

In north-eastern Brazil the water-associated disease index was applied to assess changing conditions in vulnerability over an extended time period (Dickin and Schuster-Wallace, 2014). This assessment focused on the state of Pernambuco, which is characterized by diverse climate conditions and socio-economic factors, and compared vulnerability to dengue during two time periods, 2000 and 2010. Overall the findings showed consistently higher vulnerability in humid coastal areas where most of the population is concentrated and most dengue cases are reported. However, areas currently experiencing a low burden of disease in the less developed western region of Pernambuco have experienced increased vulnerability over the study period. The results illustrate the role of environmental and social changes occurring between 2000 and 2010, such as land use intensification and growing urban population density that gradually increase exposure to dengue. At the same time, decreasing levels of susceptibility were observed in some rural areas due to improvements in access to water and healthcare over the period of analysis. While dengue prevention activities are generally focused on areas with most reported cases, such as in Recife, the state capital, these results suggest that decision-makers should additionally be aware of increasing trends in vulnerability in regions with a lower burden of disease.

The findings in Pernambuco additionally highlight interactions between temporal scales that influence vulnerability. For instance, seasonal dynamics, inter-annual droughts, and growth in the community healthcare network program over the ten-year period illustrate fast and slow processes that all influence dengue transmission, but are generally not considered together in decision-making (Srinivasan *et al.*, 2013). Tools such as early warning systems may be developed to plan for seasonal patterns, but identifying long-term changes in social and environmental conditions that increase vulnerability is also important to improve long-term planning to reduce the expansion of dengue. Interventions such as improved surveillance and healthcare system capacity to deal with dengue are needed in growing urban areas where vulnerability is expected to increase. Furthermore, common practices may need to be adjusted to adapt to changing exposure conditions, such as in the case of cistern use in the western semi-arid region of Pernambuco. Cisterns are important for conserving water during dry conditions, for reducing the economic costs of water for households, and in reducing waterborne illnesses. However, if vulnerability increases in these areas which have lower capacity to cope with dengue, cisterns must be properly sealed and cleaned to avoid unintended consequences such as creating breeding sites for mosquitoes (Regis *et al.*, 2013).

Implications for population health intervention research

The case studies in this chapter illustrate how vulnerability assessment can be applied to study population health in the context of global environmental change. Vulnerability assessment highlights important pathways where interventions can be targeted, provides information on conditions that may impact an intervention's success, and identifies areas for more detailed data collection and research. Place-based analysis is particularly valuable as vulnerable populations are heterogeneously distributed by location, and specifying a scale for analysis also leads to opportunities to involve relevant stakeholders in the process (Turner *et al.*, 2003). Importantly, place-based vulnerability should be recognized as the result of processes operating at multiple spatial and temporal scales (Adger *et al.*, 2009). Applying a vulnearbility framework that describes some of these interdependent mechanisms can promote responses that go beyond a focus on proximal actions, which are not adequate to address health in the face of global environmental change. Consideration of social and ecological systems in a vulnerability assessment further provides evidence of the importance of interventions beyond the health sector, which are increasingly important to address complex health challenges (Bowen *et al.*, 2013). For instance, reducing the need for households to store water containers in north-eastern Brazil would require steps to improve the reliability and affordability of water supplies, requiring cross-sectoral collaboration.

As the health implications associated with global environmental change are expected to increase in the future, vulnerability assessment can contribute to finding interventions to cope with and adapt to these changes. In the case of dengue, regions with high susceptibility that experience limited exposure due to low temperatures should plan for a potential expansion of the range of mosquito populations linked to climate change. The flexibility of the vulnerability mapping approach applied here permits the use of diverse datasets, such as climate change projections and socio-economic development scenarios to study these research questions further. With high uncertainty surrounding global environmental change processes, low-regret solutions are important for decision-makers. This is critical as governments may not prioritize actions to address changes that are not immediately evident or observable due to their gradual nature. By providing insight into the dynamics of social-ecological systems that determine health outcomes, vulnerability assessment can help identify "win-win" solutions which will be beneficial despite the uncertainty of future changes. For example, decisions made to reduce vulnerability can contribute towards sustainable development in addition to improving population health, and thus achieve multiple goals.

Vulnerability assessments are commonly developed with decision-makers in mind and often involve engagement steps (Turner *et al.*, 2003). Vulnerability maps can provide a comprehensive picture for decision-makers, which is useful in considering a balance between current and future needs. For instance, resources may be used to reduce susceptibility to dengue by increasing waste-collection services or improving coping capacity such as through education campaigns. While identifying important trends and places for more detailed analysis, there are challenges to disseminating findings to decision-makers, particularly in the case of maps which may be interpreted in many different ways (Preston *et al.*, 2011; de Sherbinin, 2014). Limitations and uncertainties associated with findings should be communicated, in addition to assumptions such as types of weightings used in an index methodology. An additional limitation of this approach is that even integrated vulnerability assessments must simplify social-ecological systems to conduct an empirical analysis. Not all factors contributing to vulnerability can be accounted for in an analysis, and some are difficult to empirically assess such as political and economic root causes (Cutter *et al.*, 2008). In addition, some indicators of individual health status such as immune status may not be available. However, the contributions of vulnerability assessment do not lie in predicting the extent of a particular outbreak, but in providing a holistic understanding of vulnerability in a population.

Responses to global environmental change: future research opportunities

As humans not only drive environmental changes but can take actions to positively influence the future, understanding a social-ecological system's capacity to respond to a hazard is important for decision-makers. Resilience is a concept originating from ecological thinking, but is increasingly used more generally in "sustainability science" to describe types of response to change (Turner, 2010). This broadening of the concept has led to recognition of the need to consider dynamic social change in addition to ecological outcomes, and researchers have stressed that efforts to build resilience should remain aware of competing value systems that may lead to normative definitions of "desirable" states (Cote and Nightingale, 2012). Described in this chapter as the capacity to cope and adapt to a hazard, this perspective has parallels with population health research relating to coping with stresses (Lindsay, 2003). While the water-associated disease index methodology applied in the case studies limits a deeper exploration of the links between these areas of research, a resilience perspective may provide insight into available response options for promoting population health in the face of a changing planet (Walker *et al.*, 2004). Research focused on improving the effectiveness of interventions to cope with environmental change has led to decision-oriented approaches that focus on "pathways" of change and response, rather than on the outcome itself (Wise *et al.*, 2014). This conceptualization could be applied to examine adaptive pathways through which population health interventions can be best implemented to deal with uncertain futures. Resilience has been considered not only in identifying strategies to cope with change but also in understanding the persistence of certain conditions. While high resilience is generally considered a positive condition, some researchers show that maintaining certain system structures or functions despite a disturbance may keep a population in a state of ill health (Berbés-Blázquez *et al.*, 2014). Identification of these types of relationships may be useful for decision-makers in designing interventions that facilitate transitions to a state of health.

Conclusions

Global environmental changes are occurring at an unprecedented rate, causing systemic shifts in Earth's life-supporting systems which our societies, livelihoods, and health depend on. Unlike localized environmental hazards, these changes are widespread and occurring through varied pathways and across scales. These heath challenges require urgent attention through the development of novel research designs, interdisciplinary partnerships, and interventions that respond to complex interactions between social and ecological systems (McMichael, 2006). This chapter draws upon research applying a vulnerability framework to deepen understanding of processes through which global changes are impacting population health. Applying vulnerability assessment methods offers insight into interactions occurring across spatial and temporal scales not often examined through traditional population health approaches.

The outcomes of these analyses provide opportunities for more holistic solutions, including strong justification for mitigation, better monitoring of social, economic, and physical indicators of population health, and increased cooperation between health and environment practitioners (Lindsay, 2003). Designed as tools to communicate complex relationships in a way that engages policy-makers, vulnerability assessments can identify leverage points to promote health in the most vulnerable populations. Moreover, a coordinated response is required to address interlinked processes that impact human health, such as interventions that promote health and water security through the equitable distribution of resources.

In 2015 a series of Sustainable Development Goals were negotiated in the UN General Assembly that set the stage for the future global development agenda, including a Sustainable Development Goal on improving health and well-being. As a set of integrated targets, this agenda presents an important opportunity for population health researchers to collaborate with the larger sustainable development community. As stated by McMichael (2006), population health is the "bottom line of sustainability" and intervention research is needed to find ways of improving health while addressing the uneven impacts of global environmental change.

7 The geography of malaria control in the Democratic Republic of Congo

Mark Janko and Michael Emch

Melinda Meade opened her seminal work "Medical Geography as Human Ecology" by noting that "health professionals frequently wonder how medical geography differs from epidemiology, or what geographers do that health planners do not. These are not idle questions," she wrote, "for in fact most models in medical geography have come from epidemiology or from health planners" (Meade, 1977: 379). In that paper, Meade then went on to develop her "Triangle of Human Ecology," which posits that an individual's health status is a result of interactions among his/her biological characteristics, behaviors, and environment. Since that time, modern medical geography has evolved into a field with its own identity, as well as one that continues to maintain a close relationship with epidemiology and other health science disciplines.

This chapter is motivated by the belief that geographers have a fundamental and unique role in public health research, and intervention research in particular. Specifically, we believe that geographers' main contribution to public health intervention research is through studies designed to understand how the effect of an intervention might vary across space. Methodologically, geographers' contribution is motivated by fundamentally different inferential goals than those of epidemiologists. Indeed, while epidemiology traditionally focuses on attempting to estimate a mean effect, such as an average treatment effect, medical geographers have, since Meade's writings, been encouraged to take a step back and ask whether or not such a summary measure is a reasonable one. If, for example, different habitats have varying degrees of support for a mosquito vector, why should we expect an intervention such as a bednet to have the same effect everywhere, or that the average effect would even be experienced by the majority of individuals in the majority of places? Additionally, different people have different biological characteristics as a result of living in different places, and these characteristics can act to promote or prevent disease. Furthermore, different people in different places have different behaviors, and as Meade noted, "[b]ehavior, the observable aspect of culture, usually has spatial expression" (Meade, 1977: 382). Thus, where most public health intervention research is interested in estimating an average of some sort, medical geographers' chief inferential goal should be to understand how an effect varies from place to place and what contributes to it.

Our recent work on spatial dimensions of randomized vaccine trials illustrates this point. Trials have an underlying assumption that the intervention effect is randomly distributed throughout the trial area. We recently showed this assumption is not necessarily true if one considers the spatial distribution of the trial participants, and we used such knowledge to investigate variability in the effect of an oral cholera vaccine trial using a spatially referenced database (Ali *et al.*, 2005; Ali *et al.*, 2008; Emch *et al.*, 2006; Emch *et al.*, 2007; Perez-Heydrich *et al.*, 2014). Our results illustrated that the protective efficacy of oral cholera vaccines varies in space (Emch *et al.*, 2007), and that the variation is inversely related to vaccine coverage (i.e. percentage of people vaccinated in an area) after adjusting for several ecological factors (Ali *et al.*, 2005; Emch *et al.*, 2007). We also found that higher levels of neighborhood vaccine coverage are linked to lower risk of cholera among residents (Ali *et al.*, 2005; Ali *et al.*, 2008). These findings show that higher levels of vaccine coverage can lead to higher levels of indirect protection of non-vaccinees, and may also lead to higher levels of total protection, i.e. indirect protection combined with direct protection of vaccines (Ali *et al.*, 2005). Lastly, we found that ecological factors (e.g. household wealth, proximity to aquatic reservoirs, migration) in different parts of the trial area in rural Bangladesh influenced the efficacy of the vaccine (Emch *et al.*, 2007).

There are similarities between the example of vaccine interventions for cholera control and bednets against malaria. Both are human infectious diseases with an important environmental component. Cholera is linked to the environment because the bacteria that cause the disease live in particular aquatic ecosystems. Malaria is linked to the environment because the mosquito vector has a limited environmental habitat. Both diseases occur when humans interact with their environment in particular ways and contexts. Our work here is an extension of our work on vaccine coverage. The purpose of this chapter, however, is to investigate the relationship between bednet distributions and malaria, and whether or not the effect of bednet coverage varies across communities.

Malaria transmission and the triangle of human ecology

Perhaps the word that best describes malaria transmission is heterogeneity (Figure 7.1). Indeed, transmission is highly variable both spatially and temporally (i.e. seasonally). Seasonal variation is driven by rainfall patterns, since the *Anopheles gambiae* mosquitoes that transmit the parasite most efficiently in sub-Saharan Africa breed in transient pools of sunlit, standing water. Transmission peaks shortly after the rainy season begins and continues until shortly after the rainy season ends. Additional factors include the altitude and temperature of the area, with higher temperatures and lower altitudes favoring the reproduction of *Anopheles* mosquitoes and the parasite itself, and hence favoring transmission (Afrane *et al.*, 2006; Afrane *et al.*, 2008).

Furthermore, malaria affects different populations in different ways. The most at risk population is children under five, and that risk may decline with age, based on exposure to higher, stable levels of transmission, which can lead to development of clinical immunity (Snow *et al.*, 1997; Snow *et al.*, 1999). Other population characteristics that put a person at risk include sex, with men tending to have higher risk,

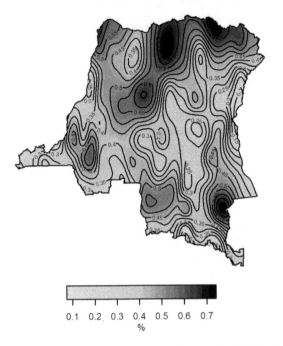

0.1   0.2   0.3   0.4   0.5   0.6   0.7
%

Figure 7.1 Surface of crude prevalence of malaria in the Democratic Republic of Congo

pregnancy, as well as sickle cell trait, which is protective against malaria (Williams *et al.*, 2005).

Behavioral characteristics also influence an individual's risk of malaria, such as occupation, with farmers being at higher risk owing to their increased time spent in habitats favored by mosquitoes. Other risk factors include whether or not individuals take anti-malarial drugs, undergo rapid diagnostic tests, or sleep under a bednet, which is the focus of this chapter.

In addition to the complexities involved in the interaction between people and their environment, malaria (and any zoonotic disease, for that matter) has the added complexity of the vector. Each of the many vector species has its own habitat preferences and behaviors, which largely go unmeasured in public health studies. Thus, it is not surprising how difficult it has been to reduce transmission. Nevertheless, the global health community has been committed to just that, and has made substantial progress.

Reducing malaria transmission

In the year 2000, approximately 2.4 billion people were at risk of malaria infection, with between 300 and 500 million of those becoming infected, leading to between 1.1 and 2.7 million deaths (WHO, 2000). At around the same time, the global community began investing considerable funds and effort in reducing malaria worldwide (RBM, 2005; The Lancet, 2006). That effort has included a suite of interventions ranging from distributions of insecticide treated bednets, indoor residual spraying, rapid diagnostic tests (RDTs), treatment with Artemisinen-based combination therapy (ACT), and intermittent preventive therapy (IPT) among pregnant women. Uptake of these interventions has been considerable, particularly in sub-Saharan Africa, which continues to have the highest burden of disease. As a result, malaria transmission and mortality have been drastically reduced. For example, in 2012, there were between 135 and 287 million estimated cases of malaria worldwide, and between 473,000 and 789,000 deaths (WHO, 2014).

Bednets

Arguably the most important intervention (or at least the most widely deployed one) in sub-Saharan Africa is bednet distributions. A considerable literature has emerged on the efficacy of bednets. To keep that discussion within the scope of this book chapter, however, we shall limit our interest to two issues: (1) whether or not the net that an individual sleeps under is untreated (UTN) or insecticide-treated (ITN); and (2) whether or not there is a community-level effect of bednet coverage on risk of malaria infection.

There is broad consensus in the literature about the efficacy of treated bednets. That consensus emerged in the 1990s, when a series of randomized controlled trials from four different areas in sub-Saharan Africa showed that intervening with ITNs to disrupt contact between mosquitoes and humans could drastically reduce transmission and mortality (Nahlen *et al.*, 2003). These four trials (D'alessandro, Olaleye, Langerock *et al.*, 1995; Binka *et al.*, 1996; Habluetzel *et al.*, 1997; Nevill *et al.*, 1996), in fact, provided the body of evidence that led the Roll Back Malaria program to adopt insecticide-treated nets as a key tool in combating malaria (Nahlen *et al.*, 2003).

One consequence of this emphasis on insecticide-treated nets, however, is that the efficacy of untreated nets was not carefully evaluated, since the global health community adopted treated nets so quickly. Indeed, while an intact net that is in good condition can provide a barrier to mosquitoes, the degree to which that barrier sufficiently reduces biting, and hence malaria infection, is somewhat unclear since a mosquito landing on an untreated net is not killed, and is thus (theoretically) free to bite when an individual emerges from the net (Clarke *et al.*, 2001). Another concern is that untreated nets might actually increase the risk of malaria for those not sleeping under any net, with mosquitoes being diverted towards those who lack bednet protection (Clarke *et al.*, 2001; Genton *et al.*, 1994; Snow *et al.*, 1988). The limited evidence that does exist, however, suggests that there is a protective effect. For example, Mwangi *et al.* (2003) found that those sleeping under an untreated net in coastal Kenya had lower odds of malaria infection than those not sleeping under a bednet, an effect that disappears if the bednet is in poor condition (Mwangi *et al.*, 2003). Clarke *et al.* (2001) present a similar finding, noting that untreated nets in good condition significantly decreased malaria prevalence in the Gambia. They also note that those not under a net did not appear to be at increased risk due to mosquitoes being diverted away from those under a net (Clarke *et al.*, 2001). D'alessandro, Olaleye, McGuire *et al.* (1995) also find a protective effect of untreated net use compared to no net use, but add that the effect is less than that of using a treated net.

While the primary interest in bednets has been on their direct role in reducing an individual's risk of malaria infection, there has been additional interest in potential herd effects of bednets. Herd effects arise when an intervention provides indirect protection to those not receiving it. These effects are most commonly seen in vaccines for infectious diseases, where individuals not vaccinated are at reduced risk of disease because those around them have been vaccinated, are protected from the infectious agent, and thus cannot transmit it on to the non-vaccinated individuals. With regard to herd effects of bednets, the hypothesis underlying such an effect is that mosquitoes landing on ITNs will be killed, thereby reducing the vector population available to bite those not sleeping under a net, and thereby reducing transmission (Magesa *et al.*, 1991). A number of examples support this reasoning. Howard *et al.* (2000) found that a protective herd effect existed up to a 1.5 kilometer radius from a household, at which point the effect began to disappear (Howard *et al.*, 2000), while Magesa *et al.* (1991) found that biting rates among individuals not sleeping under an ITN had declined by 90 percent in areas that had received net distributions.

The case of UTNs is not as clear, however, since there is not a clear mechanism in which biting would be reduced for those not sleeping under a bednet. One possible mechanism is that the increased bednet coverage within a community reduces the number of individuals who could become infected, and thus we could see a herd effect for those not sleeping under a net as a result of there being fewer infected individuals whom mosquitoes could bite and then transmit malaria on to others. Some evidence suggests that such community-level herd effects exist. Hii *et al.* (2001), for example, found that higher coverage of bednets within a short distance of an individual had a (paradoxically) stronger negative effect on that individual's odds of infection than did their personal use of a net. The same study also found that infection rates for both *P. falciparum* and *P. vivax* malaria declined with increasing bednet coverage, which led the authors to suggest that UTNs do indeed have an effect on mosquito survivability, and that, furthermore, there is no evidence that mosquitoes are diverted to biting others (Hii *et al.*, 2001).

Ultimately, the most influential studies investigating the efficacy of bednets have been, not surprisingly, randomized trials. Those studies, however, were only interested in estimating an overall average effect, whether in terms of the direct effect of a bednet or a herd effect. Furthermore, these effects differed among study areas. D'alessandro, Olaleye, Langerock *et al.* (1995), for example, noted that the effect of an ITN intervention on mortality was different across the five areas in their study, sometimes even in the opposite direction. Such a finding led the authors to state that "although studies have shown that insecticide-treated bednets are effective at preventing malaria, it cannot be assumed that control measures of this kind will be effective when introduced as a national public health measure" (D'alessandro, Olaleye, Langerock *et al.*, 1995: 482). In the case of UTNs, Hii *et al.* (2001) concluded their work by noting: "It is important to know whether UTNs have similar [herd] effects elsewhere, especially in areas of sub-Saharan Africa where *An. funestus* and the *An. gambiae* complex are the main vectors" (Hii *et al.*, 2001). Both of these statements speak to the challenge of generalizability in randomized trials, but, more importantly, to the recognition that the effects may (or do) vary across different places. To best understand efficacy, then, we must understand this place-based variability.

Understanding variability requires us to look beyond randomized trials and instead look towards large, population-based surveys conducted across entire countries. Fortunately, many diverse surveys exist to support this effort. When the global health community began to understand the effects of bednets and include them as a key tool in reducing malaria transmission, it also began leveraging such surveys to understand how the malaria burden was changing over time. Included in such surveys are a number of questions about bednets and other relevant characteristics. Working with these surveys is challenging, however. Indeed, in 2000, Eisele *et al.* reviewed the state of malaria evaluation work and recommended that more advanced statistical procedures such as multi-level modeling be employed to better understand heterogeneity in malaria transmission and, by extension, potential heterogeneity in effects of interventions. Understanding that heterogeneity, they noted, requires collaboration between medical geographers, statisticians, epidemiologists, and entomologists (Eisele *et al.*, 2000). While their recommendation was in response to a scientific challenge, data issues also pose a challenge. Since the bulk of available data is observational and comes from a variety of population-based surveys such as demographic and health surveys (DHS), multiple indicator cluster surveys, and malaria indicator surveys (Eisele *et al.*, 2000), estimation procedures must account for phenomena such as clustered data, which inflate type-I error rates and thus lead to false inferences. Given this background, we turn our attention to understanding geographic variability in the community-level effect of bednets on malaria transmission in the Democratic Republic of Congo (DRC).

Demographic and health survey and other data

The data for this study come from a variety of sources. Malaria and survey data come from the 2007 DHS that was conducted between the months of January and September in the DRC. DHSs are cross-sectional, population-based cluster surveys (http://www.measuredhs.com/). More than 260 DHSs have been conducted in over 90 low- and middle-income countries since 1984, and in many countries are conducted every 3–5 years. The DRC survey included an extensive questionnaire about a range of individual, household, and malaria-specific characteristics, and also included the collection of dried blood spots used for HIV testing. The surveys are organized by ICF International, a Washington, DC-based consulting firm, and administered in collaboration with each country's Ministry of Health. Locations of household clusters are gathered via global positioning system (GPS) receivers during data collection. The DHSs collect de-identified individual-level data that is geocoded at the household cluster level. Cluster locations are randomly shifted between 2 km (urban areas) and 5 km (rural areas) following standard DHS protocols to help protect the anonymity of survey respondents. In each survey, household clusters are randomly chosen and then weighted to become representative of the national population. The 2007 DRC DHS comprised adults aged 15–59 years living in 300 household clusters, 293 of which had GPS coordinates taken (Figure 7.2).

We acquired the leftover blood spots from the 2007 survey for all survey respondents and used them to test for malaria. We have recently published a series of papers on malaria in the DRC based on these data (Messina *et al.*, 2011; Taylor *et al.*, 2011a; Taylor *et al.*, 2011b). We used real-time Polymerase Chain Reaction (PCR) to measure malaria on 8,836 leftover dried blood spots and found that the overall prevalence of malaria parasitemia was 33.5 percent (95 percent CI 32–34.9). The cluster-level crude prevalence of malaria ranged from 0 to 88 percent and exhibited considerable heterogeneity. Previous estimates of malaria prevalence in the DRC, based upon a relatively small number (<10) of sampling sites, also indicated a high but spatially variable malarial parasite rate and risk (Gething *et al.*, 2011). Overall, coverage of anti-malarial interventions was poor in 2007. Only 7.7 percent (95 percent CI 6.8–8.6) of households with children under five owned an ITN, and only 6.8 percent (95 percent CI 6.1–7.5) of under-fives had slept under an ITN the preceding night.

Figure 7.2 Locations of 293 demographic and health survey clusters, 2007

In addition to individual and population data, we also incorporated ecological data on environmental factors affecting malaria transmission, such as altitude, rainfall, and temperature. These are used as environmental control variables. Data on altitude came from the DHS, while rainfall and temperature were measured using NASA's Tropical Rainfall Measuring Mission. Rainfall was measured for the month prior to the one in which the survey was administered in a given cluster to reflect the lag effect of precipitation on mosquito abundance. For temperature, the mean temperature during the month of the survey in each cluster was calculated.

Estimations strategy

Because spatial heterogeneity arises as a result of a number of factors related to the underlying ecology of an area, we first consider the size of an area over which that ecology is relevant to transmission. The geographic data associated with DHSs are point data, with each point representing a survey cluster – the primary sampling unit of the DHS. A survey cluster is a collection of households in either a village (in the rural case) or neighborhood (in the urban case). To link these survey clusters to relevant ecological characteristics (altitude, rainfall, temperature, etc.), we associate with each cluster an area defined by a buffer with a ten-kilometer radius, and consider this area as the community in which an individual lives and the scale at which community-level effects operate. We choose this radius because it corresponds to the maximum flight distance of a female, human blood-fed *Anopheles gambiae* mosquito, the predominant vector for malaria in sub-Saharan Africa (Kaufmann and Briegel, 2004). Additionally, given the poor conditions of the transportation network in the DRC, it is unreasonable to assume that the daily activity space of the study population is more than ten kilometers from the survey cluster. We thus limit the extent to which the interaction between population and environment can influence transmission to the ten kilometers around the survey cluster.

For inference, we proceed along two avenues. First, we employ a simple multilevel model in which we model an individual's malaria status as a function of demographic, behavioral, and ecological covariates, and allow only the model intercept to vary across the communities, thereby correcting for cluster sampling. While this approach does not allow us to understand how a community-level effect may vary between communities, it does provide us with an overall average effect of community bednet coverage on malaria transmission. Model 1 is specified as follows:

$$y_i | p_i \sim bern(p_i)$$
$$logit(p_i) = x_i^T \beta + u_j + \varepsilon_i$$
$$u_j \sim N(0, \tau_j)$$
$$\varepsilon_j \sim N(0, \tau_i)$$

Here, y_i is a vector consisting of a binary response for each individual's malaria status, which is connected to a linear predictor via a logit link. The covariates are represented by x_i^T, a 1×11 row vector of the global intercept plus other relevant factors related to transmission, including sex, occupation, age, treated bednet use, untreated bednet use, roof quality, rainfall in a community the previous month, temperature, whether or not the community is urban/rural, and community bednet use. β, then, is an 11×1 column vector estimating the average overall effect (or so-called "fixed effect") of the covariates, while u_j represents an adjustment to the global intercept and therefore accounts for the clustered sampling. Finally, ε_i is a pure noise error term. Both of the error terms are assumed to be normally distributed with mean zero.

While model 1 provides an average effect for each of the covariates, our inferential goals are in understanding how the effect of community bednet use might vary from place to place, a goal that is not met from that model. In order to gain an understanding of that variability, we then proceed to fit a multi-level model in which the community-level effect of bednet coverage is allowed to vary by cluster. Furthermore, because of the highly localized nature of transmission, we assume that the slopes will vary independently from one another. As such, we formulate model 2 as follows:

$$
\begin{aligned}
y_i | p_i &\sim bern(p_i) \\
logit(p_i) &= x_i^T \beta + u_j + \mu_j + \varepsilon_i \\
u_j &\sim N(0, \tau_j) \\
\varepsilon_j &\sim N(0, \tau_i) \\
\mu_j &\sim N(0, \tau_j)
\end{aligned}
$$

This specification is the same as in model 1, but here we have added the additional error term μ_j, which represents deviation from the mean effect of community bednet use across survey clusters. As such, estimating the entries in this 293×1 vector is what our overall inference is focused on.

From the two specifications above, we can clearly see the hierarchal nature of the models. This observation, coupled with the multi-level nature of the sampling, naturally leads us to think about implementing fully Bayesian inference by specifying prior distributions for all parameters. This has the added benefit of allowing us to fully account for uncertainty in all levels of the model, as well as make direct statements about the probability of estimated parameters given the data. For the coefficients giving us inference on covariate effects, as well as for the error terms, we specify vague normal priors with mean 0 and large variances. For the hyperparameters of the error terms (τ), our priors must come from distributions that only have support on the positive real line. The gamma distribution satisfies this requirement, and as such we specify vague gamma priors. Both models were fit in WinBUGS using the R2WinBUGS package in R (Sturtz et al., 2005).

Results

From model 1, we find that sleeping under a bednet, whether treated or untreated, has a considerable effect on an individual's log odds of having malaria.

Table 7.1 Results from model 1 (log odds)

Variable	Mean	2.5%	97.5%
Treated bednet use	−0.25	−0.53	0.002
Untreated bednet use	−0.24	−0.42	−0.07
Bednet coverage (%)	−0.01	−0.015	−0.007
DIC	8988	8859	9516

Additionally, we find that there is an added effect at the community level, whereby greater coverage within a community is associated with increased protection against malaria (see Table 7.1). Another interesting result from the model is the degree of similarity in the effects of treated and untreated net use, which may be due to: (1) greater importance of the physical barrier than the insecticide itself; or (2) UTNs are, in fact, ITNs but have not been re-treated according to the re-treatment schedule. In the latter case, this would indicate that the net itself is not necessarily untreated, but "not-retreated" and still carrying insecticide. We should also note that, while the effect of sleeping under a treated net is not "significant" in the classical sense, there is nevertheless a considerable marginal density that sits below zero. We can see this easily by way of a density plot of the posterior distribution for the effects. In fact, this is one of the benefits of Bayesian inference. Rather than a point estimate and confidence interval for a parameter, Bayesian inference provides a full probability distribution for each of the parameters in the model. We show the posterior distributions for treated, untreated, and community bednet coverage in Figure 7.3.

From model 2, we observe that by allowing the effect of bednet coverage to vary independently (in a spatial sense) across the 293 DHS survey clusters, the direct effects of sleeping under a treated or untreated net both become slightly stronger, and the effect of sleeping under a treated net even becomes significant in the classical sense (see Table 7.2 for results from the second model). In both models, the effect of treated bednets has less precision than that of untreated nets, though this may be due to far fewer treated nets being used at the time the survey was conducted. As with model 1, we provide plots of posterior densities for the effect of sleeping under a treated or untreated net (see Figure 7.4). More important than the findings on the direct effects of sleeping under a bednet is that we also find considerable variation in the effect of community bednet coverage on an individual's log odds of having malaria. Figure 7.5 shows this clearly by way of a map of the sum of the $\beta_{community\text{-}nets} + \mu_j$ terms from the second model specification, which captured varying community-level effects. In particular, we see areas with a much stronger association between community coverage and reduced log odds of malaria transmission. The strongest effects are seen in the western and south-western part of the country, which is not surprising given the increased level of access to Kinshasa and improved infrastructure allowing non-governmental organizations to more easily distribute nets. Higher levels of wealth in the area are also likely to increase net ownership. We also observe

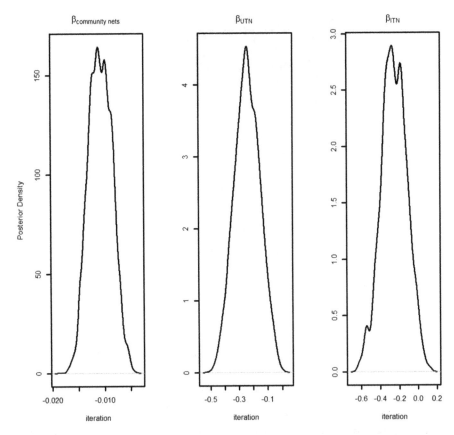

Figure 7.3 Posterior distributions and traceplots of bednet variables from model 1

Table 7.2 Results from model 2 (log odds)

Variable	Mean	2.5%	97.5%
Treated bednet use	−0.27	−0.54	−0.006
Untreated bednet use	−0.25	−0.46	−0.032
Bednet coverage (%)		See map in Figure 7.4	
DIC	8831	8770	8909

stronger effects along the southern and parts of the eastern border, where the presence of humanitarian aid has likely included bednet distributions.

In more rural areas, conversely, these effects are in the opposite direction. This finding is most notable in the far north-east and north-west of the country, as well as large swathes of the middle of the country. All of these areas are very remote and, in the case of the north-east, are experiencing pronounced and lingering

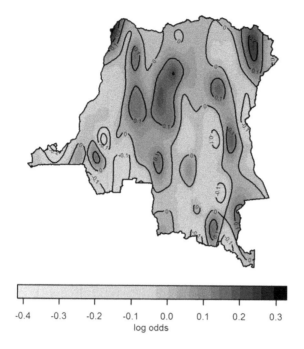

Figure 7.4 Surface of the varying effects of community bednets in the DRC

conflict. This is an interesting finding, and one that may stem from worse net quality. That potential lower quality could be due to the greater difficulty in accessing remote or conflict-ridden areas to re-treat nets. In this case, if nets are not retreated, then their herd effect will diminish, and if, at the same time, a net becomes damaged while an individual continues to use it, then a mosquito entering the net at night will be more likely to continue to bite as a result of being trapped. It is important to note, however, that *some* of these deviations, both on the strongly positive and negative side, are likely spurious, since, with 293 random slopes, it is easy to expect that some of them would be extremely positive or negative simply due to chance.

Conclusion

Billions of dollars are being invested in malaria prevention and control, and the most common intervention is bednets. Thus, understanding the effectiveness of this intervention in particular communities is essential. From our work here, we have found that a community-level effect does exist, and on average provides protection beyond that conferred by the individual-level effect of sleeping under a net. This is consistent with other work that has investigated herd effects. Unlike those other studies, however, we find that the effect varies between communities

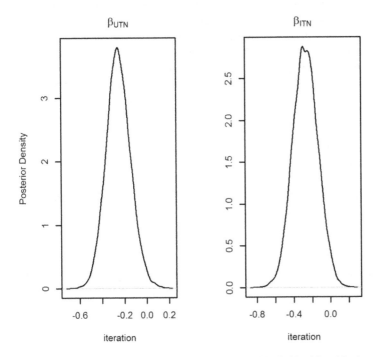

Figure 7.5 Posterior distributions and traceplots of individual-level bednet variables from model 2

and exhibits considerable geographical heterogeneity. Such a finding is important in terms of malaria control, because it provides us with an understanding of where control efforts have been stronger and have highly beneficial effects in reducing malaria incidence, and conversely, where additional resources are needed. Also, for areas where the herd effect was highly protective, it is important to investigate what factors led to this increased protection, and whether or not they can be reconstructed elsewhere. For areas where the herd effect was not strong, or trended in the opposite direction than desired, it is crucial to understand what factors were involved there as well, be they related to the local population, its behavior, infrastructure, or ecology. For example, areas with stronger protective effects tended to be in or near urban centers, such that interventions could be carried out in these communities more easily and regularly. Conversely, areas where risk increased with coverage tended to be highly rural and generally inaccessible. As a result, interventions were likely to be fewer and less frequent, resulting in communities with poorer overall net quality, such that increased coverage with poor nets could produce worse malaria outcomes. Thus, one lesson learned from this work is that achieving a high level of protection in rural, inaccessible areas requires extra resources and focused efforts. Additionally, another possible explanation for the effect we are seeing in these more rural areas is that different species of malaria-transmitting *anopheline* mosquitoes are present, and may have different behaviors such that bednets do not provide any protection. Geographical diversity in vector populations may account for some of the observed spatial heterogeneity in bednet effectiveness.

To return to where we began, Meade's triangle of human ecology posits that individuals and their behaviors interact with the environment in complex ways, leading to different health outcomes. Here, we find evidence that a public health intervention has different effects in different places, thereby influencing health outcomes in varying ways. This insight is one of the important contributions medical geography can make to public health intervention research. Our statistical work and mapping has uncovered a pattern that shows where malaria intervention effects vary, and in what direction. To proceed further, medical geographers, having observed a pattern, are compelled to try and understand the processes that produced it.

Acknowledgements

The authors wish to acknowledge funding support from NSF (BSC-1339949). Additionally, we are grateful to the Carolina Population Center and its NIH Center grant (R24 HD050924) for general support.

8 Natural experiments for reducing and preventing chronic disease

Daniel Fuller and Erin Hobin

The objectives of this chapter are threefold. First, we will define and discuss natural experiments in the context of chronic disease reduction and prevention. Natural experiment studies will be described in terms of the distinct opportunities and challenges of examining population-level or policy interventions implemented by community, government, and private-sector stakeholders. Second, we will discuss design and analysis methods to consider when conducting natural experiments. We will build on the STROBE (Strengthening the Reporting of Observation Studies in Epidemiology) and TREND (Transparent Reporting of Evaluations with Nonrandomized Designs) statements to help guide researchers conducting natural experiment studies. Finally, we will present real-world examples of natural experiments in Canada from our work. We will discuss the research and policy implications of our natural experiment studies and the challenges and opportunities they present.

Reducing and preventing chronic disease at the population level

Chronic disease contributed to a high and increasing number of deaths globally in 2013 (Lozano *et al.*, 2012). In addition to the enormous impact on mortality and quality of life, the financial costs associated with the treatment and management of chronic disease are substantial. Evidence indicates that the majority of the illnesses caused by chronic disease are preventable through behavior change including reductions in tobacco and alcohol use, and improvements in physical activity and healthy eating (Vos *et al.*, 2012). Although many efforts to address these behavioral risk factors have been implemented, most focus on individual-level behavior change (Hallal *et al.*, 2014; Heath *et al.*, 2012). These approaches show modest to large effects on behavior change among population subgroups in need of secondary or tertiary prevention, with much less success for protecting lower-risk populations from developing a disease. Given the current and increasing prevalence of chronic disease and behavioral risk factors, a combination of population- and individual-level interventions are required to achieve meaningful reductions across the population.

Individual behavior change interventions are not effective at shifting the distribution of disease risk factors at the population level for a number of reasons. First, these target high-risk individuals who are already living with chronic disease or are at high risk because of poor health behaviors (Rose, 1985). Selection based on a high level of risk is inherently secondary prevention or tertiary treatment, and does not address the underlying causes of the disease. Second, despite the efficacy of individual-level behavior change interventions for providing benefit to one individual, there is a revolving door of new individuals requiring intervention. This is demonstrated by the persistently high prevalence of chronic disease and disease risk factors in the population, including high rates of smoking, low physical activity, and poor diet, despite numerous existing individual-level interventions (Schwab and Syme, 1997). Moreover, continuous and expensive screening processes are required to identify high-risk individuals. Many people have chronic disease markers or health behaviors that place them at or near the high risk definition. Behavior change interventions focused on individuals are unable to reduce the average risk and shift the population distribution of health behaviors or chronic disease outcomes because it is not financially or practically feasible to provide the dose and resources for these individual-level interventions to reach a large proportion of the population. As the prevalence of chronic disease and behavioral risk factors become the norm rather than the exception worldwide, population-level interventions must be introduced in combination with individual-level interventions to achieve improvements in population health.

The high and increasing rates of chronic disease demand that policy-makers and practitioners implement evidence-informed primary prevention strategies at the population level. Currently, when policy-makers evaluate the most effective interventions for improving population health there is an imbalance in the existing evidence (Hawe *et al.*, 2004). Policy-makers must choose between individual behavior change interventions supported by randomized studies (e.g. randomized control trials (RCTs)), and population health interventions that are supported by non-randomized studies. When weighing the evidence, evaluation results using randomized designs are able to make stronger causal claims and are currently the preferred choice for policy-makers (Petticrew *et al.*, 2012a). Simultaneously, population health researchers frequently settle for cross-sectional designs or quasi-experimental designs to estimate the impacts of policy and environmental interventions that either lack a control group or pre-test observations. This can lead to biased or less compelling conclusions than those using more rigorous study designs.

Natural experiments

Natural experiments are interventions that can potentially shift the risk profile of the population and reduce the evidence imbalance from evaluations of individual-level focused interventions to population-level interventions. Natural experiments link researchers' questions with the answers required by policy-makers (Rychetnik *et al.*, 2002; Petticrew *et al.*, 2012a). By better aligning evidence about the impact of policies within and outside the health sector to influence health, the impact of research on policy can be maximized (Petticrew *et al.*, 2012a).

There is currently no agreed upon definition of what constitutes a natural experiment, and Table 8.1 shows a number of definitions. For the purpose of this chapter, we echo the guidance of the UK's Medical Research Council on natural experiments, which distinguishes between a natural experiment (i.e. the intervention), and a natural experiment study, which is an empirical evaluation of the impact of the event of interest on some outcome (Craig *et al.*, 2012). We understand natural experiment studies to have the objective of studying the causal effects of the event on some outcome, but where the investigator cannot assign treatment conditions and cannot assign subjects at random to treatment (Cochrane, 2008).

For example, McLaren and Emery (2012) used the natural experiment of variations in city water fluoridation policies to conduct a natural experiment study examining the impact of fluoridation on oral health in children aged 6–11 years. Using linear regression models to estimate the effect and control for confounding, the study showed that children living in cities with fluoridated water had fewer (–1.6, 95 percent confidence interval (CI): –3.4 to –0.12) decayed, missing or filled teeth compared to children living in cities without water fluoridation. In subgroup analysis the effect of fluoridation on oral health was observed in all income and education categories but was greatest for lower education and higher income households.

Exploiting natural experiments to conduct natural experiment studies is an underused tool in population health evaluation (Petticrew *et al.*, 2005). Natural experiment studies are likely part of the solution for improving the quantity and quality of evidence to better understand the implementation, impacts, and unintended consequences of interventions implemented outside of research settings. Natural experiment studies are important because they broaden the range of interventions that can be evaluated beyond those that can be planned by researchers and studied within an RCT. For example, interventions that have the potential to shift the population distribution of chronic disease exist in communities, business organizations, health systems, and governments but are rarely evaluated empirically by researchers. Expanding the scope of what can be empirically studied by health researchers by broadening the tools, skills, and accepted approaches for conducting rigorous and appropriate evaluations of policy-relevant questions begins to address the evidence imbalance.

When is a natural experiment study appropriate?

Natural experiment studies are most appropriate when four criteria are met (Craig *et al.*, 2012). The first criterion is that there are conceptual or theoretical reasons to believe the intervention will impact health, yet there is uncertainty about the effects. There are at least two plausible scenarios where this criterion is met. Natural experiments with few intervention components (e.g. a smoking ban) and large effects may provide strong evidence of health impacts and be highly policy relevant. More complex natural experiments with a large number of intervention components (e.g. a family/school education and promotion campaign) and complex interactions

Table 8.1 Natural experiment* definitions

Source	Definition
Medical Research Council	For the purposes of this guidance, we use the term natural experiment to refer to the event of interest. We use "natural experimental study" to refer to ways of evaluating interventions using unplanned variation in exposure (in the sense given above) to analyze impact.
Dunning (2008)	"Outcomes are compared across treatment and control groups, and both a priori reasoning and empirical evidence are used to validate the assertion of randomization. Thus, random or "as if" random assignment to treatment and control conditions constitutes the defining feature of a natural experiment."
Robinson *et al.* (2009)	"[N]atural experiments as opposed to random experiments imply acts of nature, or more generally, exogenous interventions demarcating observations in theoretically important ways. However, the key distinction is that the assignment mechanism is out of the control of the researcher, whereas in a controlled experiment the assignment mechanism is generated by the researcher for the experiment itself. In a natural experiment, some external force intervenes and creates comparable treatment groups in a seemingly random fashion."
Cochrane (2008)	"The objective is to study the causal effects of certain treatments or programs . . . but for one reason or another the investigator cannot impose on a subject or withhold from the subject, a treatment whose effects he desires to discover, or cannot assign subjects at random to different procedures."
Meyer (1995)	"Good natural experiments are studies in which there is a transparent exogenous source of variation in the explanatory variables that determine the treatment assignment."
Blundell and Costa Dias (2002)	"[T]he so-called natural experiment . . . typically considers the policy reform itself as an experiment and tries to find a naturally occurring control group that can mimic the properties of the control group in the properly designed experimental context."
Academy of Medical Sciences (2007)	"A natural experiment constitutes some circumstance that pulls apart variables that ordinarily go together and, by so doing, provides some sort of equivalent of the manipulations possible in an experiment deliberately undertaken by a researcher."
Last (1995)	"Naturally occurring circumstances in which subsets of the population have different levels of exposure to a supposed causal factor, in a situation resembling an actual experiment where human subjects would be randomly allocated to groups."
Shadish *et al.* (2002)	"The term natural experiment describes a naturally-occurring contrast between a treatment and a comparison condition. Often the treatments are not even potentially manipulable[.]"
Diamond and Robinson (2010)	"A technique that frequently proves fruitful in [the] historical disciplines is the so-called natural experiment or the comparative method. This approach consists of comparing – preferably quantitatively and aided by statistical analyses – different systems that are similar in many respects but that differ with respect to the factors whose influence one wishes to study."

* Because no agreed upon definition exists these definitions often interchange the terms "natural experiment" and "natural experiment study".

between intervention components may conceptually impact health but provide limited evidence of health impacts. Complex natural experiments may not provide strong causal evidence but rather should be viewed as contributing to a body of evidence and researchers should strive to ensure their studies are replicable.

Second, a natural experiment study is appropriate when it is not possible to conduct a planned experiment. There are a range of circumstances when a planned experiment (randomized or not) is impractical or unethical and a natural experiment is a reasonable alternative. For instance, it may be unethical to manipulate exposure to national legislation to study the effects on a specific health condition if it is known to have other health benefits, it has been shown to be effective in other jurisdictions, or if its main purpose is to achieve non-health outcomes. Additionally, interventions that are highly context-specific (e.g. interventions on First Nations Reserves (Lévesque *et al.*, 2005)) or unique in time and place (e.g. Olympic Games (Peel *et al.*, 2008)) are also good candidates for natural experiment studies because the intervention is not controlled by the researcher and using a randomized design is impractical or not possible.

Third, data can be obtained via surveillance or primary collection on an appropriate study population with variations in exposure to the intervention. Pre-implementation data and variations in exposure to the intervention are crucial for methodologically sound natural experiment studies. Researchers should be aware of potential natural experiment studies in their field and plan pre-implementation data collection on exposures, outcomes, and confounders. Such pretests serve many purposes, including providing data about how the groups being compared initially differ on key variables as well as expose the potential likelihood of strong initial selection biases in terms of exposures, outcomes, and confounders.

Fourth, the natural experiment has the potential for replication and generalizability. Natural experiment interventions are often designed and implemented by local actors who have the policy levers, resources, and influence required to implement novel and sustained changes at the population level. The challenge in these community-driven solutions is that they are rarely one size fits all. Instead, their design, implementation, and timing are tailored to fit the local reality of the environment. This heterogeneity across intervention sites creates significant tension between, on the one hand, acknowledging that each local context has different initial conditions, specific needs, and inherent knowledge, and on the other hand, meeting the expectation of the research community that contributions to the general body of scientific knowledge have relevance beyond the local level. Therefore, balancing fidelity to the original intervention and sensitivity to the needs of local communities and new populations requires intervention fidelity to be redefined. Rather than a standardized recognizable form that is replicated in every site, intervention fidelity could be conceived as a standardized function or purpose across sites. The intervention process and sequence eliciting change is the same in all sites, performing the same purpose, yet the form of the intervention might be different. The intervention can adapt to different initial conditions in each site while respecting scientific evidence. Replicability of natural experiments can then focus on addressing research questions comparable to past research. The questions and methods used can contribute to building the evidence base to inform the general principles of the intervention. Generalizability should focus on elements of the natural experiment with the potential for effects regardless of context. Generalizable components of the intervention should be distinguished from context-specific components based on theory or subgroup analysis (Fuller and Potvin, 2012; Petticrew *et al.*, 2012b). Such an approach to intervention research conceives complexity not within the intervention per se but in the context of the setting into which the intervention is introduced and with which the intervention interacts. This perspective allows critical components of the intervention (e.g. intervention theory) to be replicated in other settings while being interpreted and modified to fit the local context without compromising intervention fidelity.

One final note: it is not the case that when a planned experiment is not possible a natural experiment study is the alternative. A natural experiment with limited theoretical reason for a health impact or a small hypothesized effect size may not be worth the resources necessary to plan a natural experiment study. Additionally, a natural experiment study with limited variation in exposure and no appropriate comparison groups has little value over a cross-sectional study and may not add value to the literature.

Natural experiment study design and analysis recommendations

There is considerable discussion about appropriate methods for natural experiment studies. The Medical Research Council encourages the use of the STROBE (Strengthening the Reporting of Observational Studies in Epidemiology) and TREND (Transparent Reporting of Evaluations with Nonrandomized Designs) guidelines. These guidelines provide an excellent starting point for developing, conducting, and reporting on a natural experiment study. However, these current guidelines are somewhat limited in terms of their application to natural experiment studies. In Table 8.2 we suggest additional elements unique to natural experiment studies that researchers should consider and report that are not included in the original STROBE or TREND statements.

Natural experiment study examples

We provide three examples of natural experiments in Canada that we used to conduct natural experiment studies. These examples offer the opportunity to discuss the differences, similarities, and challenges of conducting natural experiment studies. The natural experiments were at the community level, at the provincial level, and within private industry.

Public bicycle share program in Montreal, Quebec

Public bicycle share programs (PBSP) are widely implemented in Western European cities, and increase population access to bicycles by making bicycles

Table 8.2 Elements to consider and report when conducting a natural experiment study

Topic	Descriptor
Natural experiment appropriate	• The intervention is highly context-specific • The intervention is unique in time and place • The natural experiment is policy relevant

RCT or quasi-design less feasible	• Impractical:
	  – Scale of intervention is too large
	  – Timeline for observable effects is too long
	• Unethical:
	  – Known risks
	  – Not accepted by community
	• Politically not feasible:
	  – Guaranteed minimum income study

Intervention effect	• Define whether a small, medium or large effect is expected based on theory and past research • Define whether the effect is expected to be immediate, lagged, or both • Define whether the intervention effect is expected to differentially impact population subgroups
Data collection	• Surveillance or routine collection of data is available. • Collection of new data specific for the intervention is possible given methodological considerations. • Confounders have been identified and measured?
Intervention	• Groups can be clearly distinguished based on variations in exposure to the intervention. • What are hypothesized mechanisms of intervention effect? • How was exposure defined? – Based on quantity and duration?
Population	• Who is the target population of the intervention? • What proportion of the population is exposed to the intervention?
Design	• What is the most closely related quasi-experimental or experimental design? • Are pre- and post-intervention collected/available? • Are treatment and control groups as similar as possible? • What methods have been used to assess similarity between groups?

available at docking stations throughout an area within a city for a fee (Shaheen *et al.*, 2010; Pucher *et al.*, 2010). For example, Montreal's BIXI (BIcycle-taXI) program was launched in May 2009 and makes available 5,050 bicycles at 405 docking stations within an area of ~46.5 km^2, encompassing ~380,000 inhabitants. Bicycles are available for a check-out fee of $5–7 for 24 hours, $40–60 for a month, or $70–100 for the season. After paying the check-out fee, the first 30 minutes of usage is free. Users extending their usage beyond 30 minutes pay a usage fee of approximately $1.50 per 30 minutes. Bicycle docking stations are on average 300 meters apart and tend to cluster around public transit hubs. The public bicycle share program was a natural experiment that had the potential to change transportation and physical activity behavior.

Research question and hypothesis

Our research team was interested in examining the impact of the PBSP on cycling, collisions between cyclists and motor vehicles, and transportation modal shift from personal motor vehicles to cycling (Fuller *et al.*, 2013a; Fuller *et al.*, 2013b; Fuller *et al.*, 2013c). We hypothesized that the PBSP would increase cycling, would not change the number of collisions between cyclists and motor vehicles, and would have a minimal impact on modal shift.

Methods

We used a repeated cross-sectional design for the study. Three population-based samples of adults participated in telephone surveys. Surveys were conducted at the launch of the PBSP in Montreal, at the end of the first season, and at the end of the second season (Table 8.3). The bike share program is available from May through November and removed from December to April because of snow. The sampling frame for each survey was individuals residing on the Island of Montreal with a landline telephone. Within households, the individual over 18 years of age to next celebrate a birthday was asked to respond. To recruit a sufficient number of respondents reporting cycling, we used stratified sampling based on the presence or absence of BIXI docking stations in the neighborhood. In the first stratum, we contacted households with a landline on the Island of Montreal using random digit dialing. In the second stratum, we oversampled households with a landline by matching postal codes provided by respondents to neighborhoods where the PBSP was available.

Source of natural variation

Our natural experiment study exploited geographic variation in exposure to docking stations to estimate the impact of the PBSP natural experiment on cycling, collisions, and modal shift. We defined exposure to docking stations using a dichotomous variable contrasting respondents with one or more docking stations within a 500 m road network buffer of their home (i.e. exposed) from those with no docking stations available within a 500 m buffer (i.e. not exposed). Road network buffers were calculated using geographic information systems (GIS). A 500 m buffer was chosen because this represents an easily walkable distance (Fuller *et al.*, 2011a; McCormack *et al.*, 2008). Unpublished sensitivity analyses using 250 m, 750 m, and 1,000 m buffers and road network distance to the nearest stations showed similar results regardless of exposure.

Results

Cycling

Our results examining whether the PBSP changed cycling behaviors showed that after controlling for selection on observable factors using statistical adjustment and selection on unobservable factors using difference in differences analysis, the likelihood of cycling for at least 10 minutes in the past week was greater for those exposed to BIXI after season 1 (OR = 1.47; 95 percent CI = 0.99, 2.19) and season 2 (OR = 2.86; 95 percent CI = 1.85, 4.42) compared with those not exposed to BIXI. The results were similar for 30 and 45 minutes of cycling per week (Fuller *et al.*, 2013a).

Collisions

Our results examining whether the PBSP increased collisions between cyclists and motor vehicles showed little evidence of a change in the likelihood of reporting a collision or near miss after implementing the PBSP (Fuller *et al.*, 2013c) PBSP users were not at a greater risk of reporting a collision (OR = 1.53, 95 percent CI: 0.77–3.02) or near miss (OR = 1.37, 95 percent CI: 0.94–1.98), although confidence intervals were wide. However, the number of days of cycling per week was associated with collisions (OR = 1.27, 95 percent CI: 1.17–1.39) and near misses (OR = 1.34, 95 percent CI: 1.26–1.42).

Modal shift

The estimated modal shift associated with the implementation of the PBSP from motor vehicle use to walking, cycling, and public transportation was 6,483 and 8,023 trips in 2009 and 2010 (Fuller *et al.*, 2013b). This change represents 0.34 percent and 0.43 percent of all motor vehicle trips in Montreal. For personal motor vehicle use, 10.1 percent of PBSP users reported replacing motor vehicle use with PBSP trips in 2010. This represents 17,078 unique PBSP users and 1,964 fewer automobile trips per day. This effect is small considering the 1,882,771 motor vehicle trips per day in Montreal.

Next steps

To address some of the limitations and additional research questions that arose during the Montreal PBSP study our research group began the International Bicycle Share Impacts on Cycling and Collisions Study (IBICCS) (Fuller *et al.*, 2014). This study used a non-equivalent groups design. Intervention cities (Montreal, Toronto, Boston, New York, and Vancouver) were matched to control cities (Chicago, Detroit, and Philadelphia) on total population, population density, cycling rates, and average yearly temperature. The study used three repeated, cross-sectional surveys in intervention and control cities in Fall 2012 (baseline), 2013 (year 1), and 2014 (year 2). With the IBICCS study we address some of the limitations of our past research, particularly the generalizability of the results to other cities. However, we recreated limitations because we used a non-probabilistic online panel rather than random digit dialing, which will likely make our sample less representative rather than more.

Table 8.3 Overview of the timing and data collection of PBSP natural experiment study

		BIXI available	Data collection	Sample size
2009	May		Pre-implementation	
	June			2,001
	July			
	August			
	September			
	October		Season 1	
	November			
	December			2,502
2010	January			
	February			
	March			
	April			
	May			
	June			
	July			
	August			
	September			
	October		Season 2	
	November			
	December			2,509

Province-wide physical education policy in Manitoba

Physical inactivity is a dominant pediatric public health concern in Canada (Katzmarzyk and Janssen, 2004; Roberts *et al.*, 2012). Recent surveillance studies suggest that up to 93 percent of Canadians aged 6–19 years do not achieve the recommended dose of physical activity required for adequate growth and health (Roberts *et al.*, 2012). Furthermore, the rates of inactive youth increase rapidly in the years following puberty and reach a pinnacle in adolescence where physical activity rates can decline by as much as 85 percent by the age of 15 (Kimm *et al.*, 2005). School physical education policies have been identified as critical for improving student physical activity levels (Fuller *et al.*, 2011b; Pabayo *et al.*, 2006) However, there has been little evaluation of such policies.

In 2008, Manitoba implemented a province-wide mandatory physical education (PE) policy in secondary schools designed to increase regular physical activity (Manitoba Education, Citizenship, and Youth, 2007). The new PE policy extends graduation requirements for all publicly funded secondary schools in Manitoba from 2 to 4 PE credits, mandating PE in grades 11 and 12 for the first time in Canada. The new grade 11 and 12 PE curriculum includes a physical activity practicum that requires a minimum 50 percent (55 hours) of the 110 credit hours focused on participation in physical activity. The 55-hour time frame for the physical activity practicum was established based on the expectation that students need to accumulate 30+ min/day of Moderate to Vigorous Physical Activity (MVPA) on at least five days per week to achieve the course credit.

Students can achieve the physical activity practicum through in-, out-, or a combination of in- and out-of-class time. The out-of-class physical activities eligible for credit include a wide variety of home, school and community possibilities tailored to meet students' needs and interests.

The province-wide adoption of mandatory PE would affect the ~65,000 adolescents attending high schools in the province and is designed to support sufficient physical activity to meet current guidelines for optimal growth and health. If delivered effectively, this would dramatically reduce the number of physically inactive youth, thereby lowering the risk of several chronic diseases, including obesity, type-2 diabetes and hypertension.

Research question and hypothesis

Collaboratively, our multi-disciplinary team were interested in examining the impact of the new school PE policy on secondary students' physical activity levels, describing longitudinal changes in, and the factors associated with the physical activity trajectories of secondary students in Manitoba in the context of the PE policy (Hobin *et al.*, 2014), and exploring stakeholder identified factors that facilitate effective policy implementation (Hobin *et al.*, 2010). Overall, we hypothesized that implementing the new PE policy would significantly increase the prevalence of grades 11 and 12 students in Manitoba who accumulate 30+ min of MVPA on most (five of seven) days of the week compared to students in a control province.

Methods

To address the three research objectives, this natural experiment study applied a pre-post quasi-experimental research design. As displayed in Table 8.4, the baseline data (T_0) were collected pre-policy implementation and preceded follow-up data collections conducted up to three time points post-policy (T_1, T_2, T_3) to permit tracking over the first three years of policy implementation.

To assess the impact of the PE policy, student survey data were collected in all publicly funded secondary schools in Manitoba (intervention condition) and Prince Edward Island (PEI) (comparison condition) pre- (T_0) and post-policy implementation (T_3) as part of an existing surveillance system. In-kind resources from governments, NGOs, schools, and school divisions in the two provinces enabled the collection of this census data. Longitudinal changes in students' physical activity were assessed using a multi-wave prospective cohort of secondary students (Grades 9–10 in 2007–8) in Manitoba. The student cohort was followed for up to four years pre- (T_0) and post-policy implementation $(T_1–T_3)$. Student participants were recruited in a two-step process. Secondary schools (n = 30) that offered grades 9 to 12, had enrolment greater than 100, and were government operated were randomly selected in blocks to best represent the urban and rural geography of Manitoba. In each selected school, a convenience sample of grade 9 or 10 PE classes was chosen for eligibility screening. All students in these PE classes were invited to participate in the study. Physical activity was measured

Table 8.4 Province-wide physical education policy in Manitoba study design and data collection timeline

		Baseline		Follow-Up		
		2008		2009	2010	2011
Objective #1						
Student survey						
	MB	T_0	POLICY			T_3
	PEI	T_0				T_3
Objective #2						
Accelerometers						
	MB	T_0	POLICY	T_1	T_2	T_3
	AB	T_0		T_1	T_2	T_3
Student survey						
	MB	T_0	POLICY	T_1	T_2	T_3
	AB	T_0		T_1	T_2	T_3
Objective #3						
School staff interviews						
	MB		POLICY	T_1		

as minutes of MVPA per day using accelerometers worn over seven consecutive days, and students' demographics were collected through a survey (see Hobin *et al.* (2014) for detailed methods). Finally, to assess the initial impression of the policy among school PE teachers, 60-minute semi-structured interviews were conducted one-year post-policy implementation with Manitoba secondary school staff from the sub-sample of 30 schools participating in the accelerometry.

Source of natural variation

The current study was designed to capitalize on the natural experiment presented by the province-wide policy mandating PE for all secondary students in Manitoba. The PE policy was implemented in all publicly funded secondary schools in Manitoba starting in September 2008. The nature and timing of the intervention was solely determined by government and education officials in Manitoba; researchers were not involved in the design or implementation of the policy. To examine the impact of the PE policy, this study compared secondary students in Manitoba exposed to the policy with secondary students in PEI not exposed to the policy. Students in PEI, the comparison province, were exposed to usual school practices including elective PE classes after grade 9, voluntary physical activity school programming such as intramurals and varsity sports, and community-based activities.

Results

Results of the survey data collected among secondary students in Manitoba and PEI are currently being analyzed. Currently, results of the accelerometer data and interviews with teachers are available and provide some insight about how the policy is working, for whom the policy is working, and under what circumstances. Consistent with previous research, results from the longitudinal cohort data confirm secondary students' physical activity declines with increasing grade levels, and the rate of decline in physical activity over secondary school is not significantly different between male and female students (Hobin *et al.*, 2014). However, the results also suggest that the decline in physical activity was attenuated among students with low and moderate baseline physical activity compared to adolescents with high baseline physical activity and among adolescents who attended schools in neighborhoods of low compared to high socio-economic status. Findings from interviews with teachers in Manitoba one year following policy implementation also indicated that the PE policy had differential affects on sub-populations of students, particularly those who were not sufficiently active before the policy and among students with access to fewer physical activity resources at school and in the community before the policy (Hobin, 2010).

Guiding Stars nutrition labeling system in supermarkets across Canada

Poor diet is now a leading risk factor for chronic disease and premature death in Canada (Lozano *et al.*, 2012). Nutrition labelling has been identified as an important tool to help consumers make more informed food purchasing decisions (Goodman *et al.*, 2011). "Guiding Stars" is a retail nutrition rating system that provides simple standardized nutrition labels on all products in a supermarket. The US-based system was adapted for Canada by an independent scientific panel with no associations with the grocery or food industry, and is administered by the Guiding Stars Licensing Company (Anon., 2015). The system is based on an algorithm that is grounded in Canada's Food Guide and recommendations from Health Canada; it generates scores for fresh and packaged food and beverage products which is translated to ratings of 0 to 3 stars. Products earning ratings of 1 to 3 stars have the stars displayed on the shelf tag beside the price. The more stars displayed on the shelf tag the higher the nutritional value of the product. With the exception of not labeling products earning a 0 star rating, the Guiding Stars system is consistent with the US Institute of Medicine's recommendations for a well-designed front-of-package nutrition label (Institute of Medicine, 2011), and thus has the potential to help consumers make more informed and healthier food choices. Loblaw Corporation Limited, the largest grocery retailer in Canada, has purchased the rights to the Guiding Stars system for Canada, and has implemented the system in Loblaw supermarkets across Canada (between August 2012 and September 2014). Loblaw is the largest supermarket retailer in Canada serving approximately 14 million shoppers per week in more than 1,000 supermarkets across Canada. Therefore, from the perspective of population interventions, Loblaw supermarkets provide a direct and frequent means to reach a large number of Canadians.

Research question and hypothesis

The primary objective of this natural experiment study was to determine the impact of the Guiding Stars system on the nutritional quality of consumer food purchases in supermarkets in Canada. We hypothesize that the nutritional quality of consumer food purchases in supermarkets with the Guiding Stars system will be higher over time compared to supermarkets yet to implement the intervention.

Methods

The natural experiment study uses a pre-post quasi-experimental research design with multiple comparison groups that receive each receive treatment at a different time (or stepped implementation) to examine the impact of the Guiding Stars nutrition labeling system. The stepped implementation (represented by the shaded areas in Table 8.5) refers to how supermarkets initially serving in the comparison groups eventually receive the intervention (Shadish and Cook, 2009), with the exception of the concurrent comparison condition (Alberta).

We collected data before and after the intervention using three data sources: (1) supermarket transaction data that is aggregated at the level of each supermarket store; (2) cardholder data providing individual level supermarket transaction data of the same customers over time; and (3) exit surveys with supermarket customers. Transaction data represent each food and beverage item sold in participating supermarkets. The supermarket transaction and cardholder data include all transactions at Loblaw supermarkets in Ontario, Nova Scotia, and Alberta from June 1, 2012 to June 1, 2015, representing time before and after the intervention. Exit surveys were conducted with consumers in supermarkets before and after the implementation of Guiding Stars, and provide estimates of consumer awareness, understanding, and use of the Guiding Stars system, as well as socio-economic and demographic information.

Table 8.5 Overview of the timing of Guiding Stars implementation

			Timing of intervention implementation		
Province	*Store banners*	*# of stores*	*Aug. 2012*	*Feb. 2013*	*Sep. 2014*
Ontario	Loblaw supermarkets	N = 44	GS	GS	GS
	All other Loblaw supermarket banners in Ontario	N = 285	X	GS	GS
Nova Scotia	All Loblaw supermarket banners	N = 24	X	X	GS
Alberta	All Loblaw supermarket banners	N = 81	X	X	X

GS = Implementation of Guiding Stars; **X** = Comparison conditions – no Guiding Stars.

Source of variation

The current study exploited the geographic variation and timing of the implementation of the Guiding Stars intervention in supermarkets across Canada. To investigate the impact of the Guiding Stars system, this research applies a pre-post quasi-experimental design with stepped implementation.

Results

Currently, only the results of the exit surveys in wave 1 and wave 2 conducted in Ontario supermarkets are available. Results of these data suggest that, while overall rates of consumer awareness and understanding of the Guiding Stars system were modest in both intervention and control supermarkets, substantially fewer consumers in intervention supermarkets reported using Guiding Stars to inform their food purchases as compared to consumers in control supermarkets. The low level of use and the gap between the levels of consumer awareness, understanding, and use, are consistent with other studies examining nutrition information on food labels (Wills *et al.*, 2009; Grunert *et al.*, 2010; Koenigstorfer and Klein, 2013) and suggests that other factors, such as food preferences or price may be important drivers of food choice over and above nutrition. Notably, most of those who were aware of Guiding Stars in the supermarket understood the intent of the system, suggesting that a lack of awareness of Guiding Stars may be of greater relative importance in preventing use of the system, compared with understanding. Consistent with the US Institute of Medicine's recommendations (Institute of Medicine, 2011), Loblaw launched a national promotional campaign in February 2015 to increase awareness of Guiding Stars – which ultimately may improve consumers' use of the system when comparing and choosing foods in supermarkets.

Challenges/opportunities with natural experiment studies

In this section we discuss the challenges and reflect on what we have learned from three natural experiment study examples. We link back to the four required elements for conducting a natural experiment study: (1) conceptual or theoretical reason to believe the intervention will impact health and uncertainty about the effects; (2) a natural experimental study is appropriate given the current research knowledge and type of intervention; (3) data can be obtained via surveillance or primary collection on an appropriate study population with variations in exposure to the intervention; and (4) the natural experiment has the potential for replication and generalizability.

Challenge: morphing designs

In the Guiding Stars and PBSP interventions the natural experiment study design changed over the course of the study. This is relatively common in natural experiment studies where researchers do not control the intervention (Ogilvie *et al.*, 2010). In the Guiding Stars study the natural experiment was based on the nature, timing, and allocation of the intervention in supermarkets across Canada. As the intervention unfolded the study changed in two important ways. First, the study lost the concurrent control condition in Alberta supermarkets, as supermarkets in Alberta received the intervention in September 2014 rather than as planned in the original implementation. Second, there was a delay in launching the national Guiding Stars promotional campaign, therefore a hypothesized change in exposure to the labels which was expected did not occur.

In the PBSP study the initial three-phase expansion period of the intervention was cancelled and all stations were implemented after the success of the initial launch phase. This meant a large increase in the number of stations and bicycles after a two-month trial period rather than a slow implementation over two years.

In both cases the change in design had implications for the hypothesized effect size of the natural experiment and the ability to control for confounders. The hypothesized effect size of the natural experiment decreased for the Guiding Stars study because of the delayed launch of the national campaign and subsequent lack of awareness of the labelling intervention, while it increased for the PBSP intervention because a greater number of bicycles and stations were implemented at a single moment in time. The Guiding Stars study's ability to control for confounding, particularly secular trends, was impacted by Alberta receiving the intervention rather than acting as a control condition during the entire study.

The lesson for researchers is to embed flexibility into the design of natural experiment studies. Though it is not possible to plan for unknown changes in the natural experiments, expect changes to occur and be creative in solving design challenges while keeping in mind that the ultimate objective is an unbiased estimate of the intervention effect.

Challenge: data availability and timelines

Two of the natural experiment study examples illustrate important challenges related to collecting data. The PBSP study in Montreal used primary data collection via random digit dialing to landline telephones which over-represent older adults and women. Sampling strategies that include cellular telephones and options to respond to the survey online would have improved the representativeness of the sample.

The Manitoba evaluation had to be mounted rapidly; there was no time to develop desirable extended baseline data specific to the PE policy. Despite the short time for study design preparation, existing surveillance systems employed in Manitoba (intervention) and PEI (control) using a survey with similar measures, conducted at similar times and among populations with similar demographic profiles (e.g. age, ethnicity), offered an appropriate mechanism for collecting baseline data on self-reported physical activity.

The lesson for researchers about data and timelines is to "keep an ear to the ground" about potential natural experiments occurring in the community while ensuring access to relevant data or having some flexible funds available for pre-implementation data collection if necessary. There have been efforts from funding agencies including the Canadian Institutes of Health Research in Canada and the National Institute of Health in the US to create funding opportunities with relatively shorter timelines to ensure pre-implementation data collection of promising natural experiments. We believe these opportunities are crucial and hope they will continue.

Opportunity: importance of partnerships

The three studies had varying levels of partnership with the organization implementing the natural experiments. In the case of the PBSP study multiple contacts were made between the research team and the PBSP organization for collaboration but no formal collaboration or data sharing agreements were reached. We believe this is the due the young PBSP organization in Montreal and a need on the part of the organization to ensure the maintenance of certain trade secrets.

The Manitoba study included a well-connected research team with health and education researchers, policy-makers, and practitioners. This team fostered rapid buy-in from schools and teachers which facilitated the process for conducting pre-post surveys among a census of publicly funded secondary schools in the province, accessing a subset of almost 500 students in Manitoba when collecting direct measures of physical activity using accelerometers at baseline and follow-up points, as well as recruiting teachers to be interviewed after the first year the policy was implemented in schools.

The Guiding Stars study had an established partnership with Loblaw which made it possible to test a retail nutrition labeling system, Guiding Stars, using a rigorous quasi-experimental design. Furthermore, the partnership enabled the use of transaction data to assess store- and individual-level trends in food purchasing behaviors, providing the most robust sources of data for measuring changes in food purchases in supermarkets. This partnership also presented a unique opportunity to evaluate the efforts of the "private sector" in health promotion.

The lesson for researchers about partnerships is to ensure ongoing and collaborative partnerships. Fostering partnerships with key stakeholders aligns with researchers' knowledge translation objectives. The time and effort required to build and maintain partnerships with various organizations is crucial for the success of individual natural experiment studies and more broadly for a research program centered around natural experiment studies.

Conclusion

The objectives of this chapter were to define and discuss natural experiments, to discuss design and analysis methods to consider when conducting natural experiment studies, and to present real-world examples of natural experiment studies in Canada. We believe natural experiment studies are crucial for increasing the evidence base for population-level interventions with the potential to reduce and prevent chronic disease. Researchers interested in natural experiment studies must seriously consider the theoretical potential of the intervention to cause changes in an outcome, make efforts to minimize bias by including control for selection on observable and unobservable factors, and carefully discuss their findings in a policy-relevant way.

9 Shaping the direction of youth health with COMPASS

A research platform for evaluating natural experiments and generating practice-based evidence in school-based prevention

Scott T. Leatherdale

Despite decades of primary prevention efforts being targeted at improving the health of Canadian youth, efforts in many domains seem to be failing as evident by the current risk behavioral profile of Canadian youth (Leatherdale and Rynard, 2013). Available evidence suggests that one of the major challenges inhibiting successful population prevention among youth in Canada was that no one was systematically collecting the necessary data to inform and evaluate prevention activities in a comprehensive or ongoing fashion. This issue was raised almost a decade ago by Roy Cameron and his colleagues specific to youth tobacco control. Cameron *et al.* (2007) highlighted that a major factor inhibiting effective school-based tobacco control prevention efforts was the inability of researchers and prevention stakeholders to generate robust information on the health of youth, to evaluate how tobacco control interventions actually impacted youth in different school contexts, and to effectively translate the limited evidence that did exist to support practical real-world action in a timely fashion. To address this ongoing challenge, there was a need to rethink how applied primary prevention research among youth populations is both *conceptualized* and *conducted*.

In terms of *conceptualized*, it is important to strengthen the relationship between the researchers that generate new evidence and the public health and school system stakeholders that are responsible for implementing prevention action (Green, 2006; Leatherdale, Brown, Carson *et al.*, 2014). In terms of *conducted*, it is recognized that studies based on cross-sectional data, and even more importantly, studies that ignore the contextual characteristics that impact the success or failure of prevention programs or policies, are not effectively advancing the evidence-base of how to improve prevention practice (Alvaro *et al.*, 2011; Leatherdale, 2012). Longitudinal data pertaining to both individual characteristics (e.g., behavioral and demographic data) and the characteristics of the contexts in which those youth are embedded (e.g. data on the program, policy, or built environment surrounding youth) are required to build a robust under-standing of how to effectively and appropriately intervene (Green, 2006; Alvaro *et al.*, 2011; Leatherdale 2012). In response, the Canadian Institutes of Health Research (CIHR) and the Public Health Agency of Canada (PHAC) are contributing to the development of this systemic capacity to link research, evaluation, policy, and practice related to population-level intervention among youth populations via their support for the development of the COMPASS system.

The primary objective of the COMPASS system is to effectively guide and continually improve youth prevention research and practice. As a learning system, COMPASS has been purposefully designed to: (1) strengthen the capacity required to plan, act, evaluate, and adapt strategies to advance youth health in multiple domains (e.g. obesity prevention, healthy diet and physical activity promotion, substance use prevention, tobacco control, bullying prevention, etc.); (2) engage researchers in studies relevant to "real-world" intervention by conducting studies that are designed to capitalize on natural experiments as a means for generating both evidence-based practice and practice-based evidence as interventions are mounted in different contexts and jurisdictions; (3) enable stakeholders in local health and education systems to plan, tailor, and evaluate local initiatives based on evidence; (4) strengthen our ability to understand and address health inequities among high-risk groups of youth (e.g. off-reserve Aboriginal youth, youth in low socio-economic (SES) communities); (5) expand our ability to understand how different social and physical environments shape youth health trajectories or impact the outcomes of different interventions over time (e.g. evaluating how different built environments can impact the success or failure of particular interventions); and (6) foster more primary prevention action and evidence-based practice through engaged partnerships between researchers and knowledge users. The COMPASS system is clearly focused on enabling the timely and robust generation of knowledge and evidence to advance youth health, by building the capacity to integrate research, evaluation, policy, and practice within the Canadian (and international) prevention system.

Why is a system like COMPASS needed?

Substance use (tobacco use, alcohol use, marijuana use) and health behaviors linked to obesity (i.e. diet, physical activity, sedentary behavior) tend to be established during adolescence with most Canadian youth exhibiting one or more of these modifiable risk factors for future chronic disease (Leatherdale and Rynard, 2013). Not surprisingly, evidence also suggests that the prevalence of these risk factors and the likelihood that they co-occur increases with age among youth populations (Leatherdale and Burkhalter, 2012; Leatherdale and Rynard, 2013). Basically, the available evidence clearly highlights that it is hard to find a student graduating from high school in Canada who is not participating in one or more modifiable risk behaviors associated with an increased likelihood of developing a future chronic disease. As such, in order to reduce the future burden associated with chronic diseases, such as cancer, cardiovascular disease (CVD), and diabetes, it is critical to promote healthier lifestyles among youth populations via more effective prevention programming.

Substance use

Age-related increases in tobacco use among youth are cause for concern as smoking is one of the major contributors to cancer, CVD, and diabetes (USDHHS, 2014). While high-risk alcohol use (referred to commonly as binge-drinking) is also associated with cancer, CVD, and diabetes (Rehm *et al.*, 2002), it is additionally associated with the three leading causes of death among youth (unintentional injury, homicide, and suicide) (Miller *et al.*, 2007). Similarly, marijuana use is associated with a variety of problems among youth populations (e.g. injuries resulting from accidents, violence associated with drug-seeking behavior, poorer academic performance) (Hall, 2009) and there is now emerging evidence of the negative long-term effects of frequent marijuana use on CVD risk (Hall and Degenhardt, 2009). Despite the risks, evidence clearly demonstrates that substance use and substance use co-occurrence remains very common among Canadian high-school students (Leatherdale and Rynard, 2013).

Obesity and correlates of obesity

Excessive weight gain among youth is associated with hypertension and abnormal glucose tolerance, and if maintained, CVD and some cancers in adulthood (Colditz *et al.*, 1996; Horton, 2009; Katzmarzyk *et al.*, 2012). The rapid increase in the prevalence of obesity suggests that the modifiable behavioral factors related to obesity (e.g. physical inactivity, unhealthy eating behaviors, sedentary behavior) may play a greater role than non-modifiable factors (e.g. genetics) (Anderson and Butcher, 2006). This is important considering that independent of contributing to obesity, physical inactivity and diet are also major contributors to cancer, CVD, and diabetes (USDHHS, 1999; Adami *et al.*, 2001; Wang, Ouyang, Liu *et al.* 2014), and recent evidence has demonstrated a link between sedentary behavior and an increased risk of future chronic disease morbidity and mortality (Tremblay *et al.*, 2010). Similar to substance use, evidence clearly demonstrates that obesity and the modifiable correlates of obesity are very common among Canadian high school students (Leatherdale and Rynard, 2013).

Why is the COMPASS system school-based?

A contextual (ecological theory) approach to population prevention recognizes that behavioral development involves relationships between the developing individual and the multiple contexts in which they are situated (Bronfenbrenner, 1979; Lerner *et al.*, 1997). The two core concepts within a contextual approach are: (1) a variety of factors from different levels of context influence behavioral development; and (2) different factors from the various levels of context coexist in a relationship in which they interact and influence each other. Since almost all youth (regardless of socio-economic status), spend ~25 hours each week in school throughout the school year, the school environment represents an important context for shaping youth behavior. More importantly, we know that the characteristics of the school environment in which students are embedded (e.g. programs, policies, built environment characteristics within or surrounding a school) can be changed to promote and/or inhibit healthier lifestyles among youth.

Research with Canadian youth has clearly demonstrated that the characteristics of the school in which a student is situated are independently associated with their likelihood of participating in a variety of modifiable risk behaviors and outcomes (Leatherdale, Brown, Carson *et al.*, 2014). For instance, we know that differences in school characteristics are associated with the likelihood that a student in Canada will: smoke (Kaai *et al.*, 2013) or be a never smoker who is susceptible to future smoking (Kaai *et al.*, 2014); use alcohol and marijuana (Costello *et al.*, 2012); be obese (Leatherdale and Papadakis, 2011); be physically inactive (Hobin *et al.*, 2012) or highly sedentary (Leatherdale *et al.*, 2010); and participate in co-occurring substance use behavior (Costello *et al.*, 2012). Although the school context is clearly an important determinant of behavior that is amenable to modification to improve student health outcomes (Bonell *et al.*, 2013), comprehensive inventories of school characteristics that may impact behavioral development (i.e. programs, policies, and built environment resources within or surrounding schools) are typically not systematically collected or examined in school-based prevention research within the Canadian context (Leatherdale, 2012; Leatherdale, Bredin and Blashill, 2014).

Our lack of available data on the contextual characteristics of the schools students attend has resulted in a very limited understanding of how to effectively change the school environment to promote health behaviors among youth. This is problematic, because despite the lack of evidence about what school-based programs or policies are actually effective, schools are increasingly being pressed to provide either 'whole-school' prevention interventions or one-off interventions to address particular prevention needs within a school. Schools are doing their best to try to promote and encourage healthy lifestyles among their students, but in many prevention domains, there is no appropriate or feasible evidence to guide them in their programming decisions (Leatherdale, 2012). There is a large gap between the type of prevention research actually being done (e.g. cross-sectional descriptive studies or randomized control trials of specific interventions) and the type of research actually needed to inform the school-based prevention agenda (Cameron *et al.*, 2007; Leatherdale, 2012).

While there is clearly some robust evidence available to guide school-based prevention programming in most behavioral domains (e.g. tobacco control, physical activity promotion, alcohol prevention), much of the available but often limited evidence is derived from artificially controlled research which does not align with the realities of 'real-world' practice within the school environment (Hawe *et al.*, 2004; Ringwalt *et al.*, 2004; Green, 2006). For instance, even if a particular school-based program or policy has demonstrated effectiveness in a randomized control trial (RCT), it may still not be effective when implemented in different school contexts (e.g. urban schools vs. rural schools). If different school contexts are not considered in the RCT, researchers would lack the insight to know if the intervention has the same effectiveness in schools where there are different built environment or community resources available (or absent) to support the intervention. For example, evidence has shown that tobacco retailer density surrounding a school is associated with youth smoking behavior (Chan and Leatherdale, 2011), but the impact of tobacco retailer density has not been controlled for in the RCTs exploring youth tobacco prevention programming (Thomas *et al.*, 2013).

Furthermore, many of the school-based interventions targeted to youth are not amenable to randomization at the school-level (i.e. policies or community built environments) given that they are dictated at a regional, provincial (state), or national level. In such instances, researchers must take advantage of quasi-experimental designs for evaluating the natural experiments that occur as school stakeholders or policy-makers implement different policies within school environments, or change community environments surrounding schools in an ongoing basis over time (Petticrew *et al.*, 2005).

Conceptual framework for COMPASS

The conceptual framework guiding the COMPASS system is adapted from my earlier research and leadership with the SHAPES model (Leatherdale *et al.*, 2009; Leatherdale, 2012), but made relevant to (1) longitudinal data systems; (2) incorporating data on the program, policy, and built environment surrounding youth; and (3) fostering knowledge exchange to promote contextually appropriate intervention action within schools. The conceptual model for COMPASS is a cyclical process that includes four staged processes and two overarching processes (see Figure 9.1). The four staged processes that occur in an ongoing cyclical order are:

1. *Data collection activities* – collecting relevant data at both the student and school levels.
2. *Knowledge translation and exchange activities* – engaging school stakeholders by providing them with (a) a timely syntheses of their school-specific data with corresponding recommendations for action and links to relevant available resources locally, and (b) COMPASS staff dedicated to working with schools to identify the prevention priorities for the school and to determine the most appropriate and feasible prevention action(s) for their particular school context and student population.
3. *Intervention activities* – mobilizing necessary staff and resources to implement the prevention action(s) identified as priorities by school stakeholders.
4. *Evaluation activities* – for all school-based prevention action(s) that occur as part of this process, COMPASS staff and researchers evaluate the impact of each intervention (both individually and comprehensively if more than one intervention were to occur in a particular school) on student outcomes to generate timely, local, practice-based evidence.

The two overarching processes that occur throughout an ongoing application of the COMPASS system are:

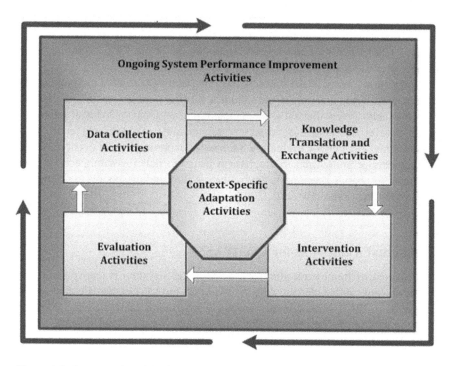

Figure 9.1 Conceptual model of the COMPASS system

1. *Ongoing system performance improvement activities* – ensuring the methods and tools used in the system are valid and reliable, maintaining data quality and improving data management techniques, continually improving and adapting the knowledge translation and exchange tools, continually updating the recommended actions, advancing the scope and quality of recommended interventions, strengthening evaluation capabilities (timing, complexity, and rigor), and strengthening the trust and relationships between system researchers and stakeholders.

2. *Context-specific adaptation activities* – COMPASS researchers work with a variety of stakeholders to establish plans for learning from the actions taken, the practice-based evidence generated from evaluating the different school-specific interventions are shared with other schools and stakeholders (especially schools with similar student populations or contextual environments where the intervention is most likely to be effective if replicated), and the research team works with stakeholders to identify new collaborators to address any knowledge gaps in the different stages of this research-to-action process.

Through this ongoing process, at a minimum, the COMPASS system has three general aims when trying to advance prevention activities targeted to youth within school settings. First, the system aims to enable researchers to work with local health and education systems to plan, tailor, target, and evaluate multiple interventions focused on multiple health and behavioral outcomes within the school context over time.

Second, the system aims to engage researchers and stakeholders in real-world studies that generate practice-based evidence about the effectiveness of interventions when implemented within different contexts and with different student populations. Evaluating the numerous ongoing natural experiments that occur within schools or communities over time can not only substantially increase the evidence-base of which interventions may be effective for which outcomes, but it also provides evidence of how different contexts can inhibit or promote intervention success (e.g. an intervention designed to remove sugar-sweetened beverages from within the school may not be effective if it is implemented in a school that is surrounded by variety stores that sell sugar-sweetened beverages), or how different student populations can inhibit or promote intervention success (e.g. the success of a smoking cessation intervention on reducing the prevalence of smokers within a school will be less likely to be effective if it is implemented in a school where there are very few smokers to begin with).

Third, the system aims to provide a platform to support and study the processes and structures required for effective knowledge exchange in school settings. Consistent with the concepts embedded in systems thinking (Leischow *et al.*, 2008), these aims are accomplished through the ongoing cycle illustrated in Figure 9.1, which links transdisciplinary action-oriented research with real-world prevention practice. To date, COMPASS system collaborators include researchers, school stakeholders, and public health system practitioners who envision a future in which schools (and the communities in which those schools are embedded) are supported by a system that enables them to pinpoint the best opportunities to improve youth health, identify effective and feasible interventions, access timely intervention resources, and use a practical data collection and feedback platform to continuously guide, evaluate, refine, and learn from their work.

Design and methodological features

The COMPASS system is founded on using rigorous data collection tools and longitudinal quasi-experimental designs to evaluate how changes in school programs, policies, and/or built environment characteristics are related to changes in multiple youth health behaviors and outcomes over time.[1] COMPASS also facilitates contextually appropriate knowledge exchange by continually providing participating schools and stakeholders with customized knowledge exchange resources to connect them to relevant prevention resources locally, provincially, or nationally.

Hierarchical longitudinal quasi-experimental design

The hierarchical nature of the design is consistent with collecting the relevant individual and contextual data required to evaluate the impact that changes to school prevention programs, policies, or built environment resources have on student behavior over time. Since student behaviors tend to cluster within schools, this creates a hierarchical structure of individual students nested within schools. The longitudinal nature of the student- and school-level data adds an additional layer of clustering within individual students and within schools over time. As such, the longitudinal nature of the clustered data in the COMPASS system requires hierarchical modeling approaches when examining the data, where longitudinal data (level 1) are nested within students (level 2) who are nested within schools (level 3) (Snijders and Bosker, 2012).

The quasi-experimental design represents a robust method for assessing causality when evaluating the impact of changes to school programs, policies, or built environment resources (*interventions*) when a randomized design is not feasible or ethical (Cook and Campbell, 1979). Using a quasi-experimental design in COMPASS allows for inferences on the relationship between a change in a school-level program, policy, or built environment resource and a corresponding change in student behavior or outcome. This method greatly surpasses other more common study designs in school-based research (i.e. cross-sectional, single cohort longitudinal) in its ability to identify causal relationships between interventions and student outcomes (Gliner *et al.*, 2009), and because it differs from the traditional experimental study design (RCT) that does not produce the evidence of effectiveness that is required for informing real-world prevention practice in school settings. Quasi-experimental evaluation of natural experiments occurring outside of a traditional experimental study design provides a means for generating practice-based evidence in school-based prevention programming by determining what works, for whom, and in what context (Petticrew *et al.*, 2005; Green, 2006), using real-world data. Although the evidence derived from a quasi-experimental design may be considered imperfect by some, it is more relevant to school stakeholders and can assist in the identification of "effective" interventions in real-world settings (Petticrew *et al.*, 2005), and it better reflects the realities of intervention implementation in real-world settings (Ramanathan *et al.*, 2008).

Incorporating linked school- and student-level longitudinal data creates databases in the COMPASS system that enable evaluation of natural experiments as school programs, policies, and resources change over time. Diffusion of innovation theory (Rogers, 2003) suggests that schools will adopt innovation at different rates; some adopt an innovation right away and some only after others have adopted the innovation. Considering that innovation adoption for schools can represent either implementing an existing evidence-based intervention or developing a new evidence-informed intervention, a system like COMPASS can take advantage of these quasi-experimental evaluations as programs, policies, and built environment resources change in schools at different times and for different reasons.

Data collection tools

The COMPASS Student Questionnaire (C_q) is used to collect the longitudinal student-level data. The C_q is designed to facilitate large-scale school-based data collections that can track individual students over time while maintaining student confidentiality. The C_q collects the longitudinal individual student-level behavioral and demographic data. The measures in the C_q can easily be adapted and changed depending on the focus of a particular project in which the COMPASS system is being used. For instance, in the COMPASS study (Leatherdale, Brown, Carson *et al.*, 2014), the C_q was designed to measure substance use, obesity, and the correlates of obesity in a brief 12-page questionnaire. It is also feasible to complement the self-reported student-level data collected by the C_q with objectively measured student-level data (e.g. genetic markers, physical measures, biological samples, etc.), depending on the study protocol and ethical approvals associated with each project or study.

The COMPASS School Programs and Policies Questionnaire (SPP), the COMPASS School Environment Application (Co-SEA), and the COMPASS Built Environment Data (C-BED) are used to collect the longitudinal school-level program, policy, and built environment data. The SPP is a paper-based survey completed by the school administrator(s) most knowledgeable about the school program and policy environment within the school. The SPP is designed to measure the presence or absence of relevant programs and/or policies, and changes to school policies, practices, or resources that relate to the different student-level outcomes measured in the C_q. The SPP is complemented with copies of the relevant policy handbook(s) or rules for each student-level outcome to facilitate additional document review if required.

The Co-SEA is a direct observation tool developed to measure aspects of the built environment within a school that are associated with obesity, eating behavior, and physical activity. As described elsewhere (Leatherdale, Bredin and Blashill *et al.*, 2014), Co-SEA is a downloadable within-school built environment scan application for use on most handheld devices with a camera. Once downloaded, Co-SEA contains an automated computer-based version of the previously validated audit measures from the ENDORSE study measuring the school food environment (Van der Horst *et al.*, 2008) and the SPEEDY study measuring the school physical activity environment (Jones *et al.*, 2010), with the additional functionality of being able to take pictures of the different within-school built environment resources identified during the audit and then storing those pictures in the corresponding data file as objective observations supporting the subjective audit measurements. This simple-to-use tool advances our ability to consistently, accurately, and quickly measure the features of the built environment within a school in COMPASS. Since it is a downloadable application, it can also be easily shared with other researchers.

The C-BED data represent measures of the built environment in the community surrounding participating schools. C-BED data are collated annually from the CanMap Route Logistics (CMRL) spatial information database and the Enhanced Points of Interest (EPOI) data resource provided by Desktop Mapping Technologies Inc. (DMTI) (DMTI Spatial Inc., 2014). The CMRL provides high-quality street map data which includes information on: street road networks; road classifications including expressways, primary and secondary highways, major roads, local roads and trails; and land-use types including residential, industrial, institutional, parks, and water bodies. The EPOI is a database of Canadian business and recreational points of interest, including: education facilities (e.g. schools and universities); golf courses; healthcare facilities (e.g. hospitals, dentists, etc.); police and fire stations; industrial facilities; food stores (e.g. grocery stores, mini-marts, alcohol and tobacco retailers); eating and drinking places (e.g. fast-food restaurants, bars); and recreation facilities (e.g. fitness centres, movie theatres). Consistent with previous research (Pouliou and Elliott, 2010; Leatherdale *et al.*, 2011), the process of identifying and linking the DMTI built environment data for schools participating in a COMPASS study in Canada involves geo-coding the address for each COMPASS school, creating circular buffers (i.e. bounded areas surrounding each school in which the different built environment characteristics are quantified[2]), selecting the built environment characteristics you are interested in including in the database, and linking the quantified built environment data for those buffers and selected indicators to each participating school. DMTI Spatial provides Canadian universities with an enterprise-wide site licence to use these built environment data for free for academic research following an initial application and, once approved, updates on these DMTI data resources are received in October or November annually.

Knowledge translation and exchange resources

Previous experience in school-based prevention programming has demonstrated that schools highly value simple context-appropriate knowledge exchange tools (e.g. customized feedback reports) that provide them with a school-specific understanding of the characteristics of their student population, the quality of the school environment they are providing to their students (programs, policies, and built environment resources), and suggestions of relevant programs, policies, or changes to the within-school built environment they can introduce within their school to improve student outcomes (Leatherdale *et al.*, 2009; Leatherdale, 2012). In order to help foster and enable COMPASS schools to create healthier school environments aligned with their own capacity and priorities, the COMPASS system currently uses three different knowledge transfer and/or exchange strategies to accelerate the translation of research evidence to real-world policy and practice. First, we have a traditional producer-push knowledge transfer mechanism via the COMPASS website (http://www.compass.uwaterloo. ca). The website provides anyone (researchers, students, school administration, parents, etc.) with a general overview the system activities and the knowledge products being produced on an ongoing basis. Second, we provide customized school-specific COMPASS School Health Profile (SHP) reports to all participating schools. The SHP allows school stakeholders to quickly and easily see, "at a glance," the risk behavior profile of their student population to inform where and how they should target future prevention activities and resources. The SHP also includes a list of evidence-based and/or evidence-informed recommendations for action (programs, policies, and/or changes to built environment resources within a school) aimed at changing the school environment to improve or maintain the health of the student population for each behavior or outcome of interest. Third, each school is assigned a knowledge broker who has specific responsibilities for working with that school to foster and advance context-appropriate prevention action within each school. In each school, the knowledge broker: (1) facilitates and enables ongoing interaction between the research team, school stakeholders, and relevant community partners; (2) assists school stakeholders in determining appropriate priorities for immediate and future action based on student needs (C_q data and the SHP results) and the available resources within a school (SPP and Co-SEA data); (3) collects process measures from school stakeholders pertaining to the different intervention actions taken within a school annually; and (4) works with school stakeholders and external partners to identify available resources (e.g. funding, physical equipment, specialized personnel) to support particular actions within a school.

Sample COMPASS system products

While the COMPASS system was only recently conceptualized and developed, there are already a few emerging examples of the system being used in practice to advance the science of school-based prevention.

The COMPASS study

The COMPASS study is a prospective cohort study designed to collect hierarchical longitudinal data from a convenience sample of 89 secondary schools in Ontario and Alberta (Canada) and the ~50,000 grade 9 to 12 students attending those schools (Leatherdale, Brown, Carson *et al.*, 2014). The COMPASS study, the first application of the COMPASS system, is among the first study of its kind internationally to create the infrastructure to robustly evaluate the impact that changes in school-level programs, policies, and built environment resources might have on multiple youth health behaviors and outcomes over time. The COMPASS study was funded by the Canadian Institutes of Health Research (CIHR).

COMPASS Guatemala

COMPASS Guatemala is a pilot study designed to translate all of the COMPASS system resources into Spanish and pilot test their use in a convenience sample of secondary schools in Guatemala. The COMPASS Guatemala study was funded by the International Development Research Centre (IDRC) as a partnership between COMPASS leadership (S. Leatherdale) and the lead stakeholder for non-communicable chronic disease prevention in Guatemala (J. Barnoya). Given that there was no research infrastructure in Guatemala to support school-based data collection and evidence-informed prevention action at the time of this project commencing (2013), the COMPASS system was seen by funders and relevant local stakeholders in Guatemala as a mechanism for rapidly and robustly building research-to-practice capacity in a low-income developing country that was lacking such infrastructure and research support.

What about SHAPES?

As mentioned earlier, the COMPASS system was adapted from the SHAPES model. SHAPES is a simple data collection and feedback system for school-based tobacco control (Cameron *et al.*, 2007) and physical activity promotion and obesity prevention (Leatherdale *et al.*, 2009). As a cross-sectional data collection and producer-push knowledge transfer platform, SHAPES was quite innovative at the time of its inception (Cameron *et al.*, 2007; Leatherdale *et al.*, 2009). However, despite the success of SHAPES as a school-based surveillance tool (primarily in the domain of tobacco control), the system was not able to generate practice-based evidence as was originally intended due to the cross-sectional nature of the student-level data and the lack of available school-level data pertaining to the built environment within participating schools. The COMPASS system corrects those deficiencies and adds additional new value in many additional domains as described previously.

Conclusion

The COMPASS system is purposefully designed to create the infrastructure required to understand how to change and adapt environmental contexts within and surrounding schools to advance youth health. The COMPASS system enables not only researchers and stakeholders to robustly evaluate the impact that changes in school-level programs, policies, and built environment resources have on multiple youth health behaviors and outcomes over time, but it is also designed to strengthen the partnership between researchers and stakeholders. Members of COMPASS include both researchers and stakeholders who envision a future in which schools and communities are supported by system models that enable them to pinpoint the best opportunities to improve youth health, identify effective and feasible intervention approaches, access timely intervention resources, and use a practical data collection and feedback platform to continuously guide, evaluate, refine, and learn from their work.

Acknowledgements

The development of the COMPASS system was supported by a bridge grant from the Canadian Institutes of Health Research (CIHR) Institute of Nutrition, Metabolism, and Diabetes (INMD) through the "Obesity – Interventions to Prevent or Treat" priority funding awards (OOP-110788; grant awarded to S. T. Leatherdale). The first application of the COMPASS system (the COMPASS study) was supported by an operating grant from the Canadian Institutes of Health Research (CIHR) Institute of Population and Public Health (IPPH)

(MOP-114875; grant awarded to S. T. Leatherdale). The second application of the COMPASS system (COMPASS Guatemala) was supported by a development grant from the International Development Research Centre (IDRC) (107467-027; grant awarded to S. T. Leatherdale). Dr Leatherdale is a Canadian Institutes of Health Research (CIHR)/Public Health Agency of Canada (PHAC) Chair in Applied Public Health. This CIHR-PHAC Research Chair supports the future capacity building of the COMPASS system. The author would also like to thank the COMPASS staff (Chad Bredin, Rachel Laxer, Audra Thompson-Haile) who have contributed to the development of the COMPASS system tools and resources.

Notes

1 Details of the design and methodology are described in Leatherdale, Brown, Carson *et al.* (2014).
2 Generally COMPASS researchers use 500 m and 1 km buffers since that is roughly the distance a student can easily walk to a return to school within one spare class or lunch period, but the size of the buffer can be determined by the researcher as appropriate for the research question being explored.

10 Exploring the implementation process of a school nutrition policy in Ontario, Canada

Using a health geography lens

Michelle M. Vine

Introduction

This research draws on the principles of health promotion and population health. Nutbeam (1996) defines health promotion as public health action intended to improve an individual's control over modifiable determinants of health such as behaviors, public policy, and living and working conditions. In this way, health (e.g. reduced morbidity and mortality, disability, quality of life) and social (e.g. equity) outcomes are the result of *healthy lifestyles* (e.g. making healthy food choices, being physically active), *health services* (e.g. prevention services), and *healthy environments* (e.g. safe physical environment, supportive social and economic conditions). Nutbeam (1996) also posits that health promotion actions such as education, social mobilization and advocacy are key determinants of health promotion outcomes, including health literacy, social action, and healthy public policy and practice. The focus of the current research is an evaluation of a particular health promoting intervention, namely a policy to support healthy consumption of food and beverages in schools.

Drawing on constructs of health promotion theory, population health interventions are developed to address a broad range of individuals by influencing the fundamental social, economic, and environmental conditions associated with health risk in an attempt to reduce health inequities (Canadian Institutes of Health Research, 2012). Interventions include programs, policies, services and resources developed within the health sector but operating outside of it (e.g. education, housing, transportation). Population health intervention research (PHIR) seeks to produce knowledge by using scientific methods to empirically explore the value and implications of interventions, the ways in which they bring about change, and the contexts in which they are most successfully implemented (Hawe and Potvin, 2009). PHIR is rooted in constructs related to the social determinants of health (Evans and Stoddart, 1994; Frank, 1995). Hawe and Potvin (2009) contend that all systematic inquiry and learning from exploring an intervention's implementation process are representative of intervention research. Thus it involves all parts of the process, including process evaluation of interventions (assessing reach, adoption, implementation, participant satisfaction) (Glasgow *et al.*, 1999), the extent to which interventions adjust to different contexts and settings, examining sustained interventions, and diffusion research (Potvin *et al.*, 2001; Steckler and Linnan, 2002).

Health geographers have become increasingly concerned with population health intervention research given that the environment plays a key role in helping to shape the conditions in which individuals experience health (Gatrell and Elliott, 2009). In geography, the relationship between the environment and human health emerged as a research focus in the 1990s. In essence, where you live (place) affects your health and well-being via access to human health risks, and the (un) availability of and access to treatment and other health services and resources (Gatrell and Elliott, 2009). Green *et al.* (2000) unpack the concept of *setting* as fundamental to health promotion theory given its role in exploring context as ecological and context-sensitive. Accordingly, the setting is defined by the subjects of intervention (individual or collective), the location of health promotion, and the setting as a target for intervention (e.g. schools, neighborhoods, etc.). While health geographers evaluate the environmental components of settings (i.e. physical, social, economic, political), social ecological theorists acknowledge that the health promotive capacity of environments should not be examined separately, but rather more broadly as a function of the cumulative impact of a range of environmental conditions on an individual's well-being over a specific time period (Kearns, 1993; Stokols, 1996). Health geographers have a clear role to play in the design, planning, and evaluation of interventions that impact the environments that support the health of the population through research that is transferrable and policy relevant, with theoretical, methodological, and substantive implications. The remainder of this chapter will focus on a case study in population health intervention research, using a health geography lens.

Childhood obesity

Obesity has emerged as a major public health issue in Canada. More than 60 percent of the adult (aged 18 or older) population is considered overweight (34 percent) or obese (26 percent), as measured by health risk associated with elevated body mass index (BMI) (Luo *et al.*, 2007; Statistics Canada, 2012). These prevalence trends translate to a tremendous burden of illness on the public healthcare system, on families, and on individuals. Direct (e.g. physician care, medications, and costs associated with adiposity and other chronic diseases) and indirect costs (e.g. the value of economic output lost due to illness, injury-related work disability, and premature death) associated with obesity are estimated to be $11 billion annually (Anis *et al.*, 2010; Janssen, 2013). Most recent estimates based on direct anthropometric measures indicate that 32 percent of children and adolescents (aged 5–17) are overweight or obese in Canada (Statistics Canada, 2012). Given that the obesity-related health outcomes for children and adolescents are varied and distinct from adults, and the strong propensity that obesity will persist into adulthood (Roberts *et al.*, 2012), it needs to be examined and responded to separately.

Obesity results from interactions between biological, behavioral, and environmental components (PHAC, 2011). While much research has been undertaken on individual-level determinants of obesity (Janssen *et al.*, 2006, for example), less has focused on the environment, particularly as a mechanism to either support or hinder opportunities for behavior modification in the areas of healthy eating (energy intake) or physical activity (energy expenditure). Obesogenic environments have been defined as the influences (barriers) that surroundings (settings), opportunities and conditions in life have on promoting energy consumption and restricting opportunities for energy expenditure at both the individual and population level (Green *et al.*, 2000; Townshend and Lake, 2009). While health geographers have made significant contributions to the literature on obesity and social and physical environments at the neighborhood level (Cummins and Macintyre, 2006; Moon *et al.*, 2007; Pearce *et al.*, 2007; Harrington and Elliott, 2009; Pouliou and Elliott, 2010), fewer have focused on the policy environment (see Asanin Dean and Elliott, 2012, for an example). Further, "[t]he growing threat that overweight and obesity poses for children of the world has been identified as a major policy issue" (Raine *et al.*, 2008: xii).

In a recent response, the Ontario government convened the Healthy Kids Panel in order to set an agenda to reduce childhood obesity by 20 percent in five years, in part by changing the food environment which they occupy (e.g. establish a school nutrition program for all publicly funded Ontario schools; develop a single standard guideline for food and beverages sold where children learn) (Ontario Ministry of Health and Long-Term Care, 2013). The remainder of this chapter will focus on the school nutrition policy environment in Ontario. In this way, this environment is explored in an attempt to understand how policies and guidelines act as pathways through which the environment shapes opportunities for energy intake (food consumption) in youth.

The intervention

In September 2011 the Ministry of Education in the province of Ontario implemented a new School Food and Beverage Policy (P/PM 150) (Government of Ontario, 2010) that would be implemented across all (72) school boards in the province. In conjunction with the province's commitment to make schools healthier, the objective of P/PM 150 was to help improve educational, attitudinal (e.g. food preferences and eating behaviors) and health-related outcomes (e.g. reducing the risk of students developing chronic diseases: type 2 diabetes, cancer, and heart disease) via nutritional standards (i.e. sell most, sell less, not permitted for sale). P/PM 150 guidelines allow schools to have ten special-event days in which food and beverages are exempt from nutrition standards. Other requirements include: that schools comply with the Trans Fat Standards set out in Ontario Regulation 200/08 and Regulation 562 related to preparing, serving, and storing food and beverages; that students are ensured access to drinking water during the school day; that schools develop strategies to reduce the risk of exposure to anaphylactic causative agents; and that schools consider the diversity of students and staff through the accommodation of religious and/or cultural needs (Government of Ontario, 2010). P/PM 150 also recommends that boards avoid offering food and beverages as an incentive to students and, if available and possible, sell food and beverages produced in Ontario.

School nutrition policy research

Given that they are where children spend much of their waking time and where they are considered to be a captive audience, schools are an important environment where teachers can support and provide leadership for positive behavioral changes. Schools have therefore been the setting for focused health promotion interventions, particularly in the area of nutrition (Parcel *et al.*, 2000). In a seminal review paper on school nutrition policies in Canada and the US, the findings of McKenna (2010) highlight that despite the existence of a limited number of evaluations, school policies can lead to a positive impact on food availability and student nutrient intake, and that behaviorally focused nutrition education may impact the eating habits of students. However, these findings also point to gaps in research in the areas of food marketing in schools, providing nutrition-related services in schools, and coordinating food services (McKenna, 2010). In Canada, only a modest amount of research and evaluation exists in the area of nutrition policy and programming in school settings (McKenna, 2010; Vine and Elliott, 2014a).

For example, in British Columbia, Mâsse *et al.* (2013) found that implementation of daily physical activity and food and beverage policy guidelines are advantageous when compared with the status quo, are well matched with teaching philosophy and school mission, and had positive impacts, but are also considered to impede schools when they are complex to implement and understand. Implications include monitoring policy implementation, providing support to schools during implementation, and impacts of guidelines on student behaviors (Mâsse *et al.*, 2013). Also in British Columbia, the findings of Watts *et al.* (2014) illustrate reduced access to sugar-sweetened beverages, French fries, chocolate and candy, salty snacks, and baked goods. Positive changes to the school food environment were linked with the expectation of school boards to implement policy guidelines (Watts *et al.*, 2014). The increased availability of specific healthy food and beverages in British Columbia schools is linked to local-level nutrition resources, capacity, and practices (Mâsse and de Niet, 2013).

In Alberta, findings illustrated the importance of identifying a school champion as a key step in adopting health promotion interventions (Quintanilha *et al.*, 2013; Downs *et al.*, 2011). In addition, several barriers were identified, including parents' resistance to change and cost, student preferences, lack of knowledge, the schools' physical location, and the provision of healthful food (Downs *et al.*, 2012).

In Nova Scotia, the results of a recent population-level study examining the implementation of a provincial policy (Fung *et al.*, 2013) indicate a positive influence on energy intake, healthy beverage consumption, and nutritional quality, in addition to the need for more interventions beyond schools. Mullaly and others (2010) in Prince Edward Island found a significant decrease in low-nutrient density foods and an increase in vegetables and fruits following the implementation of a province-wide school nutrition policy.

The existing literature provides only a partial understanding of the context of school nutrition policy and programming in Canada. Most Canadian research has occurred in Alberta, British Columbia and Prince Edward Island; a lack of research has been undertaken in this area in Ontario. Nearly all of the existing research has taken place in the elementary school setting, with virtually no Canadian data from the secondary school setting where youth have increased control over food-related decision-making, there are opportunities for youth to purchase food offsite (e.g. fast-food restaurants, variety stores), and youth are more likely to engage in risk-taking behaviors (e.g. alcohol, smoking, drug use). Several studies have observed the health outcomes associated with school nutrition interventions (e.g. weight status, consumption), the availability of food in school environments, and the success of nutrition education. Little research has examined specific contextual factors influencing school nutrition policy implementation at the local level.

The case study presented in this chapter explores the extent to which policies and guidelines act as environmental pathways through which opportunities are shaped for energy intake in youth. In this way, its focus is on the school nutrition policy environment in Ontario, Canada in order to investigate local-level factors shaping implementation. The research was guided by the following objectives: (1) to evaluate the consistency of policies across a range of spatial contexts (i.e. national, provincial, regional); (2) to examine the perceptions of key stakeholders involved in policy implementation; and (3) to investigate the perceptions of the user group.

Analytical and theoretical frameworks

This research drew on a population health framework, which acknowledges the role of both individual (i.e. biological and behavioral) and environmental determinants of health, operating outside of the health system (Evans and Stoddart, 1994; Frank, 1995). Health outcomes are impacted by interactions between multiple determinants of health: components of the social environment (e.g. education, income, culture); aspects of the physical environment (e.g. urban design, access to water); genetics; and individual behavior (Kindig and Stoddart, 2003). The complexity of these interactions was explored to understand the extent to which environmental factors shape the implementation of a school nutrition policy in Ontario, Canada.

The Analysis Grid for Environments Linked to Obesity (ANGELO) framework (see Figure 10.1) was adopted in order to facilitate and extend this examination (Swinburn *et al.*, 1999). The ANGELO framework was developed to identify and conceptualize contributions resulting in an obesogenic environment. From an ecological perspective, obesity is defined as "a normal response to an abnormal environment rather than vice versa" (Egger and Swinburn, 1997: 477), and acknowledges the need for a focus on the environment as a key pathway to human health.

A two-by-four grid, ANGELO separates environments into distinct levels (micro or macro) and types (economic, socio-cultural, physical, and political). Micro-environments refer to settings where individuals engage and interact based

SCALE ⟍ TYPE	Micro-environment (settings)		Macro-environment (sectors)	
	Diet	*Physical Activity*	*Diet*	*Physical Activity*
Physical		What is available?		
Economic		What are the financial factors?		
Political		What are the rules?		
Socio-cultural		What are the attitudes, beliefs, perceptions and values?		

Adapted from Swinburn, Egger and Raza, 1999

Figure 10.1 The Analysis Grid for Environments Linked to Obesity

on common objectives and close geographical proximity (e.g. schools, work-places). Macro-environments function at a higher level, influencing micro-environments (e.g. health and education systems). The environments in which they occur shape important determinants of health and obesity, including physical activity and food intake (see Figure 10.1). The economic environment includes the cost of food, and policies and interventions that impact food intake (food subsidies, food cost). The physical environment is represented by "what is available," including nutrition education and training opportunities. The political environment refers to laws, policies and institutional rules guiding food-related behaviors of individuals. The socio-cultural environmental involves the values, culture, and attitudes inherent within a community or society (Swinburn *et al.*, 1999).

Research methods

Mixed methods were adopted in order to meet each of the research objectives (each research objective corresponds to a specific phase of research) (see Table 10.1).

In order to explore the consistency of policies across a range of spatial contexts (research objective 1), in *phase one*, relevant English language policy documents and technical reports across Canada were selected at two broad levels (national, n = 8; provincial, n = 24), and were published between 1989 and 2009. At the time of this research, school nutrition policies had been developed in only five of the ten Canadian provinces.

Regional (n = 26) level policies published between 1989 and 2011 were selected from Ontario.[1] Two categories of documents were included: (1) policies representing nutrition standards and criteria as related to food offered for sale in schools; and (2) technical reports including healthy guidelines and research conducted in schools, published by government and non-governmental agencies. Documents

Table 10.1 Research study design

Research objectives	Method	Sample	Analysis
1. To explore the consistency of policies across a range of spatial contexts (i.e. federal, provincial, regional)	Documentary analysis	*Policy documents/ technical reports (1989–2009)* Federal (N = 8) Provincial (n = 24) *Policies (1989–2011)* Regional (n = 26)	Qualitative thematic Quantitative
2. To examine the perceptions of key stakeholders involved in policy implementation	Semi-structured interviews	Key stakeholders (n = 22)	Qualitative thematic
3. To investigate the perceptions of the user group	Focus groups (n = 3)	Secondary school students (n = 20)	Qualitative thematic

were publicly available and were included if they addressed two of the following: (1) a school-based setting (primary and secondary schools); (2) a population focus (children and youth between two and 17 years); (3) school nutrition (food for sale in school tuck shops, cafeteria, etc.), and guidelines for school nutrition; (4) the health status of students (healthy living, overweight/obesity); and (5) population health (interventions, nutrition programs). Documents were excluded if they had a focus on non-school settings or an exclusive focus on school-based physical activity (for additional methodological detail see Vine and Elliott, 2014a).

In order to examine the perceptions of key stakeholders involved in policy implementation, in *phase two*, in-depth, semi-structured interviews were undertaken – in person or via the telephone – with key stakeholders between December 2011 and March 2012. They lasted a maximum of 60 minutes. Stakeholders were recruited from local agencies supporting school nutrition programming (local public health unit and community agencies supporting school nutrition) (n = 8), and secondary school principals, vice principals, and teachers (n = 14) from nine secondary schools across three Ontario school boards (Vine and Elliott, 2014b). An interview guide was informed by the ANGELO framework, relevant literature, and the research objectives (Swinburn *et al.*, 1999). Interviews were tape-recorded (with written permission) and transcribed verbatim for subsequent thematic analysis.

Key interview topics were relevant to: regional school nutrition; stakeholder perceptions of the school nutrition policy or program operating in their respective school, or the focus of organizational activities, including facilitators and barriers associated with implementation, strategies for improvements, and perceptions of user satisfaction.

In order to investigate the implementation of the Ontario School Food and Beverage Policy (P/PM 150) (Government of Ontario, 2010) from the perspective of the user group, in *phase three*, three focus groups were undertaken (lasting between 45 and 60 minutes in length) with secondary students (n = 20) in two school boards representing both high- and low-income neighborhoods in fall 2012 (for additional methodological detail see: Vine *et al.*, 2014). Key focus group topics included: healthy eating (perceptions, importance, how the types of food eaten influences health, weight, school performance, etc.); healthy eating at school (food purchasing at school, access to off-campus food outlets, nutrition policy/ programming); and determinants of healthy eating (factors impacting food decisions, changes to school food environment). An external transcription firm transcribed digitally recorded (with participant permission) interviews verbatim for subsequent thematic analysis (Vine *et al.*, 2014).

Data analysis

In *phase one*, qualitative and quantitative methods were adopted to analyze data. First, thematic qualitative analysis was undertaken to explore high-level themes within documents operating at the federal and provincial levels (i.e. principles and objectives, key stakeholders, system themes). A coding template was designed and tested on a sub-sample (n = 5) of randomly selected documents. Second, regional-level policies (n = 26) were compared and contrasted in order to assess the extent of their consistency with new Ontario school nutrition guidelines – Policy and Program Memorandum 150 (P/PM 150) (Government of Ontario, 2010). In this way, an SPSS file was developed based on the key themes generated (i.e. school nutrition standards and criteria, anaphylaxis policy, rationale linking nutrition and learning, implementation and monitoring, food labels, food preparation, and diversity of students and staff) during qualitative analysis (according to whether or not they included each theme (yes/no)). The objective was to understand if themes generated during qualitative analysis of federal and provincial documents translated to policies operating at the regional level.

In *phase two*, interviews were digitally recorded and transcribed verbatim for subsequent analysis. Data analysis occurred concurrently with data collection, a process that helped to shape ongoing data collection, including as it related to question refinement (the ability to pursue other avenues of inquiry) (Pope *et al.*, 2000). A coding template was developed based on analyses of a sub-set (n = 5) of randomly selected transcripts. Four environmental components of the ANGELO framework (Swinburn *et al.*, 1999) were mapped onto the data, resulting in the emergence of sub-themes. An inter-rater reliability exercise was undertaken by two researchers, resulting in an agreement score of 64 percent (Miles and Huberman, 1994).

As in phase two, in *phase three*, data analysis and interpretation of key themes occurred concurrently with data collection, in an iterative process from which within- and across-group similarities and differences emerged (Miles and Huberman, 1994). A coding template was developed in this phase as it was in phase two.

In all three phases, qualitative data were coded by hand in order to assign passages of text to individual codes. In each phase, data were then uploaded to NVivo software (8.0) for further thematic analysis. Key themes emerged deductively as they related to research objectives and inductively from the documents. Results are presented according to the key themes identified by the ANGELO framework: economic, physical, political, and socio-cultural environments (Swinburn *et al.*, 1999). Phase one is presented first in an attempt to provide an overview of the broad landscape of school nutrition in Canada and more narrowly in Ontario.

Results

Phase one

Results from phase one indicate that most documents were published in Canada between 2004 and 2009, and, further, more were published in Ontario than in any other province (see Table 10.2). Ninety-two percent of regional-level school nutrition policies were published during a seven-year period (2004–11). In phase one, results are provided as they relate to the physical and socio-cultural environments, given that the Comprehensive School Health (CSH) model identified the importance of these two environments in school settings (PHAC, 2013), and the lack of data related to socio-cultural factors present in obesogenic environments (Asanin Dean and Elliott, 2012).

Table 10.2 Spatial and temporal distribution of school nutrition policies and technical reports in Canada

Spatial \ Temporal	1989–91	1992–4	1995–7	1998–2001	2001–3	2004–6	2007–9	2010–11	TOTAL
Ontario- regional[a]						***	***** ******	***** *****	26
British Columbia							† *		2
Alberta						†	†		2
Saskatchewan						††	† †		4
Manitoba						†	†		2
Ontario						† † † *	†		5
Quebec						†	†		2
New Brunswick	*		†				*		3
Prince Edward Island				†					1
Nova Scotia							*		1
Newfoundland					†		*		2
Canada	†		††		†	†††	†		8
Total	**2**	**0**	**2**	**1**	**3**	**18**	**20**	**10**	**58**

* Signifies a policy document; † signifies a technical report.
[a] Implementation data were unavailable for two regional-level school nutrition policies.
Adapted from Vine and Elliott (2014a).

Physical environment

In the context of the physical environment, the results illustrate that federal- and provincial-level documents were consistent in their messaging around the availability of nutritious food and marketing to children. A key federal priority included accessing nutritious food; nutrition education and student food choice were high priorities at the provincial level (see Table 10.3). Food availability lends itself directly to food choice in schools, workplaces, and grocery stores. The school setting was a specific focus at the provincial level, including in classrooms, cafeterias, and tuck shops. Nutrition education was a high priority at both federal and provincial levels, particularly in the context of curriculum improvements. These findings provide evidence that policy-makers are acknowledging the role of education to inform food-related decision-making. Given the extent to which the accessibility of nutritious food was a key priority at the federal level, school personnel can play a role in increasing the awareness of students about how, where, and when they can access these foods in the school environment.

Socio-cultural environment

Within the socio-cultural environment, results indicate that a health promotion framework guides a large proportion of federal- and provincial-level documents, including as a way to influence school food culture and initiatives aiming to reduce rates of overweight and obesity and related risk factors. An increased examination of local level factors impacting the environments in which health conditions are shaped is needed. A key provincial-level priority centers on the need for nutrition frameworks and intervention targeting the school and

Table 10.3 Physical and socio-cultural components of the school nutrition environment

Micro-environment	Three key themes	% of sources (# sources)
Socio-cultural	**Federal**	
	Health promotion framework	100 (8)
	Multi-sectoral approach to policy implementation	100 (8)
	Marketing to children	100 (8)
	Provincial	
	Health promotion framework	79.1 (19)
	Partnership approach to policy implementation	58.3 (14)
	Marketing to children	45.8 (11)
Physical	**Federal**	
	Availability of nutritious food	75.0 (6)
	Nutrition education	75.0 (6)
	Accessibility of nutritious food	75.0 (6)
	Provincial	
	Availability of nutritious food	66.6 (16)
	Nutrition education	66.6 (16)
	Student food choice	50.0 (12)

Adapted from Vine and Elliott (2014a).

community settings (CSH, programs). Key stakeholders include parents and families, students, all levels of government, non-profit agencies, and private sectors. Multi-sectoral collaborations and partnerships are needed across government, the private sector, and non-profit agencies in an attempt to leverage resources to support policy implementation. Improving marketing to children was also an important priority at federal and provincial levels (Vine and Elliott, 2014a).

Results reveal that more than 92 percent of policies were guided by a link between student nutrition and learning. Sixty-five percent of policies contained guidelines related to an anaphylaxis policy, while an equal proportion of policies included information about how often to sell specific food products (sell most 80 percent of the time; sell less 20 percent of the time; not permitted for sale at school). Fewer policies (62 percent) referred to the implementation and monitoring of a school nutrition policy.

The objective of phase one was to understand if themes identified in federal- and provincial-level policies and technical reports translated to regional-level policies. While the availability of nutritious food was a consistent theme across federal- and regional-level policies, it translated directly to regional level policies via nutrition criteria (i.e. fat and sodium content) and nutrition standards (i.e. sell most, sell less, not permitted for sale) where food is offered for sale in the school setting. Therefore higher-level policies directly inform school-level policies. Access to nutritious food in schools, although a key federal priority, was less of a provincial priority and absent from regional level policy. Finally, as above, nutrition education (i.e. curriculum, nutrition initiatives, programs) was a federal- and provincial-level priority. While results indicate that most regional policies in the sample (92 percent) include references to the relationship between nutrition and improved learning, few strategies were noted to improve the quality and amount of nutrition education in schools (Vine and Elliott, 2014a).

Phases two and three

Themes are presented according to their alignment with the ANGELO framework (Swinburn *et al.*, 1999).

Economic environment

Results from key stakeholders and members of the student user group indicate a link between P/PM 150, the cost of nutritious food for sale, and reduced revenue generation at the school level. Healthy food is more expensive than non-healthy food, therefore policy-compliant food for sale in the school cafeteria often acts as a barrier, particularly for vulnerable students. This barrier was pronounced even more so in schools containing a larger low-income student population. Strict nutrition policies appeared to force some students to off-campus fast-food outlets or variety stores, where food is less costly and often less nutritious. As a result, given that schools receive a portion of funding from sales, a reduction in sales translates to lower school-level revenue and a need for additional fundraising by schools (Vine and Elliott, 2014b). As per P/PM 150 guidelines, which limit food-related fundraising to ten days per school year, opportunities to augment funding lost through reduced cafeteria sales are also limited. Schools are, therefore, at risk of losing their cafeterias altogether.

Physical environment

Respondents were in agreement that the geographical proximity of schools to fast-food restaurants acted as a barrier to school nutrition policy implementation. For students attending schools in or near a downtown core, unhealthy, inexpensive foods were more likely to be in walking distance, and they were therefore more inclined to purchase these types of foods. Unrestricted access to off-campus food outlets of this nature may be linked to a loss in cafeteria revenue, leading to perceived competition between internal (i.e. vending machines, tuck shops, school cafeterias), adjoining (e.g. recreation centre attached to the school), and external food outlets. While the link between healthy eating and student learning has been established in the literature and in the first phase of this research, the role of hunger acted as a barrier to student learning. In this way, many schools operating snack and meal programs (e.g. subsidized food offered to students) were guided by the connection between healthy eating and student learning in an effort to adhere to nutrition standards, and change food preferences and eating behaviors. Students compared themselves to others and felt a sense of stigma if they could not afford to purchase food during school hours – either within or outside of the school premises. Student respondents also acknowledged the need for improved cafeteria space so that all students had the option to eat together in one place (Vine *et al.*, 2014). This was seen as an important part of a supportive school nutrition environment.

Political environment

A range of respondents highlighted the restrictive nature of the school food and beverage policy (i.e. nutrition standards), with many acknowledging that secondary students should be provided with the option to make their own food-related choices. Many students indicated that they would have liked to be engaged in the process of policy development and ongoing nutrition promotion at the school level. A number of key stakeholders were noted as being integral to supporting successful school nutrition policy implementation. They include school-level personnel, school healthy action teams, government ministries, parents, students, members of the student council, and cafeteria providers, all of whom could partner in their efforts. Although some respondents felt that school-level personnel should be responsible for school nutrition, others felt that the primary responsibility is with parents (Vine and Elliott, 2014b). The role of public health in school nutrition policy implementation and related initiatives, however, deserves further attention.

Socio-cultural environment

The school nutrition culture in schools largely relies on the buy-in and ongoing support of key school-level staff members, particularly in the context of promoting and facilitating food purchasing for student nutrition programs. Many student respondents were highly aware of the link between body image, self-esteem, mental health, and healthy eating, including the impact of healthy food on health and well-being. Providing a choice in the types of food offered for sale in schools was perceived as one strategy to engage students. For example, one school hosted a school-based advocacy team to develop an organic food garden, which provided food for school meals and informed student education for crop sustainability and local eating. One secondary school participated in a culinary arts program that involved students in food preparation for the student breakfast program, helping them to develop and extend their culinary skills.

Implications for research, policy, and practice

Given the varied landscape of school guidelines in Canada, in addition to the fact that a national school meal program does not exist, all school-level cafeterias are therefore driven by revenue generation. An examination of the implementation process of P/PM 150 provides useful contextual data in order to better understand the extent to which the intervention is being implemented as planned (i.e. intervention fidelity). As such, the perceived role that the P/PM 150 policy intervention has in raising the cost of nutritious food in schools means that fewer students purchase policy-compliant food in cafeterias, and more students purchase non-policy-compliant food off-campus. This change in purchasing behavior lends itself to students purchasing less-nutritious foods off-campus and a potential loss in school cafeteria revenue. Future analyses of the external food environment surrounding schools would help to facilitate and extend this research, particularly in the context of health-related policy. In addition, future research evaluating policy implementation 4–5 years post-implementation would provide additional data regarding uptake and success that could be complemented by data related to behavioral outcomes (e.g. healthy food intake, external food purchasing behaviors, etc.).

From a health policy perspective, the findings demonstrate the clear role that environments play in shaping (by either hindering or facilitating successful implementation) the implementation of population health interventions. For example, the socio-cultural environment of schools (i.e. cafeteria layout, role modeling) can act as a key barrier to food policies if they do not support implementation. There is an important opportunity for schools to heed messages about how to create school environments that can help to support child and youth health behaviors, in part by making it easier for students to understand the health and social implications of healthy food consumption, the importance of eating with other students in a food-friendly setting, and the role teachers and administrators can play in modeling healthy eating and food purchasing in the school setting.

Student involvement in nutrition programming (i.e. breakfast, snack and lunch programs) as both users and via student advisory councils can act as an enabler of successful policy implementation, including through peer role modeling, and interacting with administrators and teachers responsible for supporting the school nutrition environment. The value in student garden programming was a key finding, and acknowledged as a valuable way to engage students in food and food security issues. Further, engaging a broad range of multi-sectoral stakeholders in school nutrition policy implementation through regular communication strategies during policy implementation would help to enhance alignment and consistency across schools and school boards. This study represents an important substantive contribution to the limited Canadian literature on the implications of non-funded school nutrition policies and how their implementation was conceptualized.

This case study illustrates the very important role for health geographers in population health intervention research given that the environment helps to shape the conditions in which individuals experience health (Gatrell and Elliott, 2009). Health geographers – not unlike social ecological theorists – explore environmental components of intervention settings as a function of their health promotive capacity, and in doing so are well situated to support the design, planning and evaluation of population health interventions (Kearns, 1993; Stokols, 1996).

Acknowledgements

This chapter is dedicated, with love, to Dad (Bryan Vine) and Mom (Nancy Fenton).

Note

1 "The decision to focus on two different publication time periods for national and provincial (1989–2009) and regional (1989–2011) documents reflects the fact that 81% of regional level policies emerged between 2007 and 2011, subsequent to the implementation of several key national and provincial technical reports and policies related to the vital role of nutrition in the school setting (between 2004 and 2006) in Canada" (Vine and Elliott, 2014a).

11 Food retail environments in Canada

Evidence, framing, and promising interventions to improve population diet

Leia M. Minaker, Catherine L. Mah, and Brian E. Cook

In Canada, dietary risks comprise the largest burden of disease, when expressed as a percentage of disability adjusted life years (Institute for Health Metrics and Evaluation, 2010). This includes poor quality diets characterized by low fruit and vegetable intake (Hung *et al.*, 2004; World Health Organization and Food and Agricultural Organization of the United Nations, 2004; Dauchet *et al.*, 2009; Boeing *et al.*, 2012), excess energy-dense, nutrient poor foods high in fat, sugar, and sodium (Garriguet, 2004, 2007, 2009), and high intakes of "ultra-processed" foods (Monteiro, 2009; Moubarac *et al.*, 2012; Monteiro *et al.*, 2013). Individuals who adhere to Canada's Food Guide to Healthy Eating recommendations are more likely to meet nutrient requirements (St. John *et al.*, 2008), but this comprises only 0.5 percent of Canadians (Garriguet, 2009). Most Canadians eat far fewer than the Guide's recommended seven to ten servings of fruits and vegetables each day (Garriguet, 2004, 2007) In terms of diet-related chronic disease, the global obesity epidemic that has been identified as a public health crisis by Canadian governments (Public Health Agency of Canada (PHAC), 2010; Healthy Kids Panel, 2013) and international agencies (World Health Organization (WHO), 2012b) is increasingly attributed to caloric overconsumption rather than inadequate energy expenditure (Shelley, 2012; Swinburn *et al.*, 2011).

Over the past several decades, the understanding of individuals' diets has evolved among researchers from the traditional notion of individual dietary "choices" to a broader, more complex ecological understanding of diet, which recognizes the importance of multi-level influences at the individual, interpersonal, organizational, community, and public policy levels (Richard *et al.*, 2011). In the past few years, Canadian federal, provincial, and territorial governments have also begun to publicly support the notion that diet and obesity occur within complex contexts, and that addressing these issues will require Canadians to have access to safe, acceptable, affordable, and nutritious foods (PHAC, 2010; Health Canada, 2013; Healthy Kids Panel, 2013; Minaker, 2013).

The existence of key structural drivers for dietary change, including systems-level economic and social conditions, suggests that structural solutions could have a major impact on population-level dietary outcomes. Policy interventions at the population level are one way to create the healthy and supportive environments needed for dietary improvement, which is a core health promotion activity

(WHO, 1986). Importantly, health promotion is also intrinsically concerned with health equity as a foundational component of public health policy and practice. Therefore a key underlying motivation for public health to lead interventions to improve diets and health behaviors at the population level is to mitigate underlying social, economic, and spatial disparities, to promote a fair distribution of resources, and to enable individual skills and capacities.

The concept of the *food environment* captures these multi-level influences and is increasingly recognized as a significant determinant of people's diet and health over and above individual preferences, knowledge, and behaviors (Hill and Peters, 1998; Story *et al.*, Neumark-Sztainer and French, 2002; Story *et al.*, 2008; Lytle, 2009). For the purposes of this chapter, the *food environment* is defined as the availability, accessibility, and adequacy of food in a community or region (Minaker, 2013). Although inadequate income is the most significant barrier to healthy food purchasing, many individuals and families also live in neighborhoods that have few quality food retail options within easy walking distance, poor access to transport, and increased access to unhealthy foods (Fielding and Simon, 2011; Minaker, 2013). Food environments have been identified as a key predictor of people's ability to eat well and have consequences for overall population health and well-being. Therefore food environments have become an important area for population health intervention research. A challenge for population health intervention research in this area is that the vast majority of food environment interventions remain un- or under-evaluated. A series of recent reviews have highlighted specific concerns about methodology and research gaps. Two major gaps in the field of food environment research include over-reliance on cross-sectional study designs and inconsistent use of both food environment assessment methods and dietary assessment methods (Van der Horst *et al.*, 2007; Holsten, 2009; Feng *et al.*, 2010; Caspi *et al.*, 2012; Giskes *et al.*, 2011; Ni Mhurchu *et al.*, 2013; Engler-Stringer, Shah, Bell, *et al.*, 2014; Kirkpatrick *et al.*, 2014).

In this chapter, we discuss healthy food retail interventions as one particular type of food environment intervention aiming to create healthy and supportive eating environments, particularly among populations at risk of developing nutrition-related chronic disease. First, we present a conceptual understanding of food environments. Second, we discuss implications of framing food environment "problems." Third, we review evidence on the impact of healthy food retail interventions on diet-related health outcomes with a focus on healthy corner stores as a promising practice.

A conceptual understanding of food environments

The food environment includes community features related to geographic food access, such as the number and kinds of food outlets in people's neighborhoods. It also features the consumer experience, such as the kinds of foods that are available, affordable, and of good quality in existing food stores; organizational settings where people spend time purchasing food and eating it; and the information environment (Glanz *et al.*, 2005). Many jurisdictions in Canada have begun to act on food environments with the aim of creating places that are more health-promoting (Minaker, 2013). Within a given jurisdiction, any food environment action must consider the overarching governance context for policy; the physical and spatial features of the built environment; and the social and cultural community context. Recently, two reviews have both examined the socio-demographic patterning of food environment features in Canada and identified Canadian studies examining food environment features and associations with diet-related health outcomes (Minaker, 2013; Black, 2014). First, we begin with a discussion outlining current conceptual understandings of food environments.

An ecological model of food environments

The most widely used conceptual model of food environment factors that incorporates the broad policy context, specific environmental components, and dietary outcomes in a parsimonious description remains the "community nutrition environments" model by Glanz *et al.* (2005) (see Figure 11.1). This is the model that we focus on most closely in this chapter.

The community nutrition environments conceptual model is based on a socio-ecological model of health and is intended as a starting point for categorizing and thinking about environmental variables related to eating behaviors. The Glanz model incorporates constructs theoretically and empirically related to eating patterns from several academic fields, including public health, health psychology, consumer psychology, and urban planning, and is notable for distinguishing constructs that tended to be aggregated in previous research. Glanz *et al.* (2005) characterize food environments as comprised of four dimensions: the community nutrition environment; the consumer nutrition environment; the organizational nutrition environment; and the information environment (Glanz *et al.*, 2005).

The *community nutrition environment* is reflected in measures of geographic food access like the number and kinds of food outlets in people's neighborhoods. Methods of measuring geographic food access include measuring the proximity of homes to grocery stores or fast food outlets, or counting the number of convenience stores within a specific geographic area and calculating the percentage of "healthy" food retailers as a proportion of all food retailers in a given region.

The *consumer nutrition environment* represents characteristics of the food environment important to consumers who have already reached their food store or restaurant destinations (e.g. food availability, affordability, quality, and on-site barriers to and facilitators of healthy eating). The consumer nutrition environment is roughly understood as the "consumer experience" dimension of food environments. Multiple methods of measuring different aspects of the consumer nutrition environment have been developed. For example, measuring the cost of a nutritious food basket in a given geographic area, assessing the shelf-space of fruits and vegetables relative to energy-dense snack foods, or assessing the quality of fresh produce have all been proposed as measures of the consumer nutrition environment (Minaker, 2013).

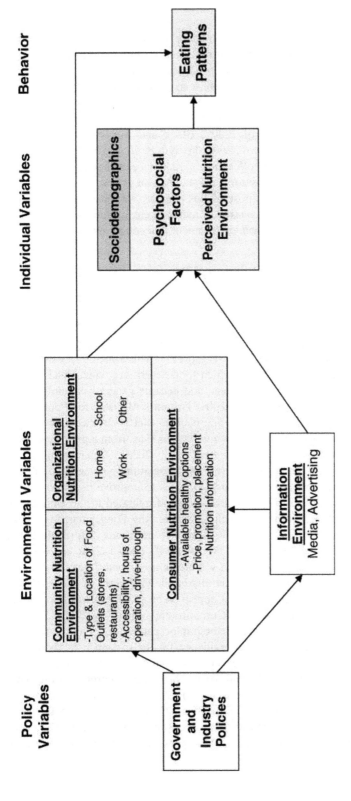

Figure 11.1 Socio-ecological model for food environments

Source: Glanz et al. (2005).

Glanz *et al.* also distinguished the broader information environment of media and marketing and the organizational nutritional environment, where food environment factors influencing health are structured within institutional settings. Finally, socio-demographic factors were seen as mediating and/or moderating the impact of food environment variables on eating patterns.

Other conceptual models of the food environment have also been published (see, for example, Papas *et al.*, 2007; Black and Macinko, 2008; Ford and Dzewaltowski, 2008; Story *et al.*, 2008; Forsyth *et al.*, 2010; Rose *et al.*, 2010). We focus here on Glanz and colleagues' model since, as noted, it has been used by a large number of food environment studies and is both theoretically driven and parsimonious. For the purpose of this chapter, we focus on the *community* nutrition environment and the *consumer* nutrition environment as constructs that together comprise the retail food environment in a community.

Framing the food environment problem

The phrase *food desert* first took hold in Britain in the 1990s as a way of describing the problem of access to healthy foods for low-income residents of poor neighborhoods in British cities (Wrigley, 2002). Food deserts are generally understood to be deprived or marginalized areas where affordable, nutritious foods are unavailable (Fielding and Simon, 2011; Minaker, 2013). Despite a lack of clear evidence on the extent to which food deserts are widespread in Britain, food deserts entered the policy debate and became generally assumed to exist (Wrigley, 2002). For example, in 2001, Britain's Minister for Public Health, Yvette Cooper, stated, "Ministers' approach to diet is focused now on issues such as how low-income families living two bus rides from a grocery get access to fresh fruit and vegetables" (Wrigley, 2002: 2032). As emerging research began to show inconsistent patterns of association between neighborhood income and geographic access to food, policies to improve deprived neighborhoods' access to nutritious foods continued and migrated to the United States, where current federal initiatives such as the Healthy Food Financing Initiative aim to support projects that increase access to nutritious, affordable foods in under-served communities (Policy Link, 2014). As early as 2006, a review of the emerging literature suggested that in some countries, food deserts may be more pervasive than others, with the United States showing signs of widespread food deserts whereas food deserts did not appear to be as widespread in other countries for which data were available (Cummins and Macintyre, 2006). Two recent reviews have examined the socio-demographic patterning of food environment features in Canada (i.e. examined evidence for the existence of food deserts) (Minaker, 2013; Black, 2014). Both reviews found that food deserts are generally not widespread in Canada, and in fact that access to grocery stores (which are sources of affordable, nutritious foods) does not appear to be consistently patterned by area-level socio-economic status (Minaker, 2013; Black, 2014). In fact, several studies conducted in urban areas have found that more disadvantaged neighborhoods actually have *better* access to supermarkets than wealthier neighborhoods (Apparicio *et al.*, 2007; Black *et al.*, 2011; Mercille *et al.*, 2013; Polsky *et al.*, 2014).

More recently, the concept of *food swamps* has been gaining traction (Fielding and Simon, 2011; Minaker, 2013). Similar to the concept of food deserts, the concept of food swamps focuses on marginalized or deprived areas, but instead of grocery stores, food swamps are related to typical sources of energy-dense foods such as fast-food outlets and convenience stores (Fielding and Simon, 2011). Of the 13 Canadian studies that have examined food swamps (i.e. neighborhood socio-economic factors in relation to geographic access to fast-food outlets or convenience stores), ten found evidence for the existence of food swamps in Edmonton (Hemphill *et al.*, 2008; Smoyer-Tomic *et al.*, 2008; Wang, Qiu, and Swallow, 2014), south-western Ontario (Polsky *et al.*, 2014), Saskatoon (Engler-Stringer, Le, Gerrard, *et al.*, 2014), the province of British Columbia (Black and Day, 2012), Montreal (Daniel *et al.*, 2009; Kestens and Daniel, 2010), Quebec City (Drouin *et al.*, 2009), and Hamilton (Latham and Moffat, 2007)). Three found mixed results (in a national school-based sample (Seliske *et al.*, 2009) the province of Nova Scotia (Jones *et al.*, 2009), and Montreal (Mercille *et al.*, 2013).

The distinction between food *deserts* and food *swamps* is an important one, reflecting the importance of framing in policy processes and the development and implementation of interventions to address problems at the population level (Schön and Rein, 1994; van Hulst and Yanow, 2016). Frames define problems, diagnose the causes of problems, imply moral judgments about problems, and also suggest resolutions to the problems identified (Entman, 1993). Thinking through the problem of food deserts requires an understanding of the definition: disadvantaged areas with *insufficient access to sources of healthy food*. On the other hand, food swamps are disadvantaged areas with *too much access to sources of unhealthy food*. The policy or program response to a food desert is much different to a policy or program response to a food swamp. A food desert can be "fixed" by opening a grocery store or fruit and vegetable market in an under-served, low-income area. Policies aimed at "fixing" food deserts are common in the United States (for example, the Healthy Food Financing Initiative (Policy Link, 2014) as previously noted). Conversely, a food swamp can be "fixed" by limiting the number of fast-food outlets or convenience stores within a geographic area, or by changing existing food retail to increase the availability of nutritious, affordable foods. Policies aimed at "fixing" food swamps have been implemented in Los Angeles (Sturm and Cohen, 2009), where a zoning regulation banning new fast-food establishments for one year was passed unanimously by the city council in 2008. Several municipalities in Quebec are currently considering zoning regulations to prohibit convenience stores or fast-food outlets from within walking distance of schools (Minaker, 2013), and the Quebec Public Health Association (Association pour la santé publique du Québec, 2011) and Québec en Forme (Québec en Forme, 2012) have produced reports to help municipalities move forward in using zoning to improve food environments around schools. Using zoning to restrict certain types of retailers from operating, however, is politically less popular than opening or incentivising grocery stores in under-served areas. Some critics have complained of "Nanny State Nonsense" in response to discussions of using zoning to ban certain types of outlets, since it interferes with their sense of liberty (Fernandez, 2006).

The way food environment problems are framed impacts the solutions that are proposed, as described above. Frames also impact selection of food environment assessment measures at the front end of the research. Currently, over 500 food environment measures exist (National Cancer Institute, 2014). Measures should be theoretically chosen to assess the types of factors in the food environment thought to be in the causal chain of dietary intake. For example, Lytle (2009) outlines one such theoretical causal chain to be the availability of low-fat milk relative to high-fat milk in the store (the food environment measure), which influences the purchases of low-fat milk relative to high-fat milk, which influences the consumption of low-fat milk relative to high-fat milk, which influences the fat content of the diet of store customers, which ultimately influences population-level disease related to fat in the diet (Lytle, 2009). Table 11.1 shows several diverse types of food environment measures, example indicators, and potential implications to outline how the choice of measure (often implicitly) frames the potential implications. In short, the type of measure chosen opens up different conversations about what can be done to solve the "problem." The next section discusses one promising "solution" to the problem of food swamps: healthy food retail interventions.

Healthy corner stores: a promising model of healthy food retail intervention

Population health interventions have been defined as "actions with a coherent objective to bring about and produce identifiable outcomes . . . [including] policy, regulatory initiatives, single strategy projects or multi-component programmes" with the intended goal of improving the health of communities or populations (Rychetnik *et al.*, 2002: 119). Hawe and colleagues have offered guidance on how to appraise the evidence on complex interventions (Rychetnik *et al.*, 2002), as well as how to design population health intervention research (Hawe and Potvin, 2009; Hawe *et al.*, 2012).

This section reviews the population health intervention research evidence base on the impact of healthy food retail interventions on diet-related health outcomes. We highlight the example of retail interventions in small food stores (e.g. healthy corner stores programs) as a promising practice.

Most of the food retail interventions that have been evaluated have been conducted in the United States, but several jurisdictional examples have also emerged in Canada (Minaker, 2013). Healthy food retail interventions usually fall into one of two types: (1) novel food retail, introducing food retail where none currently exists (e.g. opening a grocery store in a low-income, under-served area); or (2) food retail adaptations, changing existing food retail to encourage the purchasing and consumption of nutritious food, such as the "healthy corner store" approach.

Table 11.1 Food environment assessment techniques, indicators, and potential implications

Technique	Example indicators	Potential implications
Map of "healthy" food outlets	"X % of residents have access to a grocery store within walking distance of home."	If the community is dissatisfied with the percentage of residents who have access to a grocery store within walking distance, the community may want to create incentives for grocery stores to open in under-served areas.
Map of "unhealthy" food outlets	"There is an average of X fast-food outlets within walking distance of high schools in our community."	If the community is dissatisfied with the number of fast-food outlets within walking distance of high schools, they may create incentives for healthy, quick-serve restaurants to open near schools or may explore using zoning to restrict access to fast-food outlets around schools.
Shelf-space measures	"There is X times as much shelf-space dedicated to energy-dense snack food as there is fruits and vegetables within walking distance of residents' homes."	The community, in partnership with local stakeholders and retailers, could undertake programs to increase the availability of fruits and vegetables or to decrease the availability of energy-dense snack foods.
Nutritious food affordability measures	"Healthier versions of the same food product are an average of X% more expensive than their less healthy counterparts."	Communities could explore pricing mechanisms to improve the affordability of nutritious foods relative to less nutritious options.
Nutritious food affordability measures	"It costs $X for a family of four to eat a healthy diet per month in our community, which is Y% of social assistance payments in a month."	Food basket costing measures have been used to describe the difficulty of families on social assistance to afford nutritious foods.
Food availability inventory measures	"In low-income neighborhoods, X% of residents were within walking distance of a store that sold adequate varieties of nutritious foods. In contrast, Y% of high-income neighborhood residents were within walking distance of a store that sold adequate varieties of nutritious foods."	Food inventory measures are used to describe the variety of different nutritious foods available within a store. Communities could explore introducing sources of additional varieties of nutritious foods (e.g. mobile good food trucks, temporary farmers' markets) into under-served neighborhoods.
Food quality inventory measures	"In rural areas, X% of available varieties of produce were of high quality, compared to Y% in urban areas."	In communities where food quality is of concern, diverse groups of stakeholders can collaborate on improving distribution chains to ensure the availability of high-quality produce in all areas.

Different policy levers can be used to encourage each of these types of retail interventions. For example, New York City's FRESH (Food Retail Expansion to Support Health) program (New York City, 2013) entails a collaboration between the city's health, planning, and economic development departments to offer store owners and developers both zoning incentives (development rights, reduction in parking requirements) and financial incentives (real-estate tax reductions, sales tax exemptions for facilities renovations) to promote grocery store retention and development in under-served neighborhoods (New York City, 2013). Another intervention that introduces new retail in undeserved areas is the mobile vending model. Recent examples in urban environments have included New York's green carts, Toronto's Mobile Good Food Market, and Chicago's Fresh Moves bus. These "mobile" options usually entail reinterpretation of licensing, vending, and transportation policy instruments to expand retail availability of fresh produce in under-served neighborhoods outside of a bricks-and-mortar store (FoodShare, 2013; Tester *et al.*, 2010).

Healthy corner stores models, on the other hand, tap into the existing small food retail market and encourage health-promoting adaptations to that business model. Convenience stores are one of North America's largest and most diverse retail sectors and food is increasingly becoming an important revenue source for convenience stores (Agriculture and Agri-Food Canada, 2010). Start-up capital for healthier corner stores has become increasingly available, including through public funds, such as US federal or state government healthy food financing initiatives (e.g. under the auspices of the "Let's Move" childhood obesity reduction initiative) (Healthy Corner Stores Network, 2013). Healthy corner store programs can rely on less intrusive policy instruments – i.e. healthier foods are promoted as a viable market alternative rather than restricting the availability of unhealthier options (a substitution model).

The 2013 *Working With Grocers* report by Health Canada documented the most common types of store modifications that comprise healthy corner store interventions: visual promotion of healthier options, including point of purchase information programs (including displays of print material, electronic media, grocery store tours, taste tests and cooking demonstrations, sometimes with a registered dietitian); economic incentives for purchasing nutritious foods through price reductions and coupons; and increased stocking of nutritious items to increase availability and prominence of fruits and vegetables (Health Canada, 2013).

Effective programs tend to be those that are based on a behavior change theory (meaning that a full spectrum of environmental influences on behavior is taken into account, as well as the potential for local variations in the way different contexts influence individuals); are of sufficient duration to establish the presence of change; include take-away reminders for home use; are based on clear goals and objectives; use highly visible, targeted messages in information and promotional material; and are multi-faceted.

The sustainability of food retail programs depends heavily upon capacity building and financial feasibility. For example, changing a traditional corner store to a "healthy food retail" model potentially requires small-scale retail entrepreneurs to invest in major infrastructure modifications such as display shelving and refrigeration units, store reorganization, and novel process flows, to be able to provide perishable, nutritious foods (e.g. fruits and vegetables). In addition, store owners selling perishable items are required to have food handling certification, which requires additional training and cost to the store owners. Finally, procuring perishable, nutritious foods and selling them at a reasonable cost may be perceived as a barrier to owners and operators who may lack capacity (e.g. a truck) to procure foods at a wholesale price.

In Canada, several examples of healthy food retail programs exist at the city-region level, although most have taken place at a provincial or federal level.

In Saskatoon, the Good Food Junction is a good example of novel food retail. It is a cooperatively owned full-range grocery store in a low-income, under-served neighborhood (Good Food Junction, 2014). The Good Food Junction opened in 2012 and is a social enterprise aiming to provide good food at fair prices to core neighborhood residents while increasing community-based economic and social development (Good Food Junction, 2014). An evaluation of the impact of the Good Food Junction on core neighborhood residents' dietary outcomes is currently underway (personal communication, Rachel Engler-Stringer, December 16, 2013).

Among food retail adaptations, the Zhiiwaapenewin Akino'Maagewin program in First Nations communities in Northwest Ontario evaluated a comprehensive diabetes prevention program (Ho *et al.*, 2008; Rosecrans *et al.*, 2008). One of the arms of the multi-component intervention was to improve availability and afford-ability of nutritious foods in local stores, which complemented the other interven-tion components, including a school curriculum intervention. Currently, the first government-funded Healthy Corner Store evaluation in Canada is being under-taken in a low-income neighborhood in Toronto, Ontario, led by Toronto Public Health.

The state of the evidence on healthy food retail interventions

The population health intervention research evidence base has examined the impact of interventions in both larger and smaller food retail stores. To date, there is limited evidence on the effectiveness of supermarket interventions on customer purchasing behavior. In one review, eight of 13 reviewed studies collecting supermarket sales data found an increase in targeted product purchases as a result of the intervention. Point-of-purchase interventions seemed to be more effective in improving nutrition and dietary practices when paired with other strategies, such as promotion and advertising, increased availability of healthy food, and pricing (Escaron *et al.*, 2013). This review, overall, found that combining demand- and supply-side strategies significantly influenced customers as well as store owner/operators towards more healthy food purchases. The same review noted that mass media campaigns accompanying point-of-purchase interventions have been effective population-level strategies to change consumers' low-fat beverage purchasing.

The literature on small food store interventions, such as healthy corner stores, has been more promising, demonstrating a greater potential to change diet at the population level. Healthy corner store programs appear to be a promising practice for three reasons. First, small food retail interventions have had a greater likelihood of success relative to food retail interventions in larger stores. Second, while small food retailers face many business risks, the feasibility of making adaptations or changes to their retailing and marketing practices is often greater, relative to larger retailers. Third, convenience stores are often highly geographically accessible, meaning that the reach of small food store interventions may be proportionally greater.

One recent review evaluated ten studies that examined the impact of small food store interventions on consumer purchasing and consumption (Gittelsohn *et al.*, 2012). Of these, nine observed significantly increased purchasing frequency of at least one promoted food. No significant body mass index changes were found in any of the four trials that examined this, perhaps because the trials were too short to document a change in body weight.

Another recent review assessed 13 studies to date that examined the impact of interventions in outlets offering ready-to-eat prepared food (Gittelsohn *et al.*, 2013). Studies employed diverse outcome metrics, including awareness of interventions, frequency of purchasing healthy, promoted foods, and store sales data. In general, results were promising, showing that cost-effective methods (e.g. labeling foods as healthy) may have a significant impact on prepared-food sales and consumer behavior (Gittelsohn *et al.*, 2013).

As far as we know, all food retail interventions in small stores to date have focused on low-income populations, and most have also targeted racial and ethnic minority communities (Gittelsohn *et al.*, 2012). This is in contrast to studies examining prepared food; in a recent review by Gittelsohn *et al.* (2013), the majority of research was conducted in predominantly white areas and only three of 13 reviewed studies reported targeting low-income areas. In communities with limited access to healthy foods, combining culturally sensitive demand- and supply-side strategies is effective in promoting positive food-related behaviors (Escaron *et al.*, 2013).

Research on food retail interventions is growing, but much work still needs to be done. A few key gaps in the literature bear mention. First, few risks or unanticipated consequences of food retail interventions have been identified in the literature to date. The risk of profit-loss for the food store owner is one potential challenge (Gittelsohn *et al.*, 2012) that could be addressed through formative evaluation of the intervention. Formative evaluations are useful for enabling adaptations during implementation.

Second, numerous limitations in the research to date are related to study design. The limited use of randomization in existing food store intervention studies reflects the inherent difficulties of applying experimental designs to community-based health promotion and the greater suitability of quasi-experimental designs (Escaron *et al.*, 2013; Gittelsohn *et al.*, 2013). Few small-store intervention studies have considered sales data or consumer impact outcomes, such as diet or health (Gittelsohn *et al.*, 2012). In prepared-food source interventions, studies tend to lack comparison groups, selection criteria are not readily apparent, and there is a lack of adequate measures of consumer impact (Gittelsohn *et al.*, 2013).

Third, more evidence providing contextual analyses of feasibility, acceptability, and the policy context is needed. Food store interventions seem to be feasible (Escaron *et al.*, 2013) although one of the major omissions (in prepared-food source interventions in particular) is the lack of formative research, which would increase feasibility (Gittelsohn *et al.*, 2013). To engage food store owners and managers, practical strategies that will change consumer behavior should be accompanied by a return on investment for increasing access to more healthy foods (Gittelsohn *et al.*, 2014). Combining strategies to reduce unhealthy food stocking and consumption and training to reduce profit-loss risks should be included in future programs to increase sustainability. Policy options to modify retail food environments include mandates or licensing requirements for healthy food stocking. Indeed, efforts can be made to translate current small-store intervention findings into policy. Such policies should consider zoning or licensing mandates, economic incentives (coupons, produce coolers, tax breaks); improved store facade or layout; and incentivized partnerships between producers, manufacturers, and distributors (Gittelsohn *et al.*, 2012).

Conclusions

This chapter has presented healthy food retail interventions as one type of food environment intervention aiming to create healthy and supportive eating environments. Conceptual models of food environments were outlined, and the implications of framing food environment "problems" were explained. Finally, evidence on the impact of healthy food retail interventions on diet-related health outcomes was described.

While we have identified healthy food retail interventions (in particular, healthy corner stores) as a promising practice, population-wide dietary outcomes with structural causes cannot be isolated from the economic constraints that individuals and households face in acquiring healthy food of good quality in socially appropriate ways, and in our diverse Canadian context, culturally acceptable foods as well. Although the problem of addressing adequate income at the individual and household level and its direct and indirect effects on health is important, it was beyond the scope of this chapter. Future research and action on healthy food retail interventions should include diverse stakeholders and be evaluated through a variety of metrics to ensure that anticipated and unanticipated outcomes are noted.

Public health practitioners and population health intervention researchers should consider how long-term, multi-sectoral, and multi-agency networks could address economic development in low-income areas with low food availability and high rates of chronic disease (Gittelsohn *et al.*, 2012). Public health practitioners should present up-to-date evidence when approaching grocery store owners or managers to participate in interventions. In many prepared-food source interventions to date, public health actors have taken a leadership role in imple-mentation. Partnerships between public health and academic institutions may overcome many of the aforementioned evidence gaps through improved social marketing of program strategies and benefits as well as rigorous evaluation, which would include impact assessments using psycho-social surveys and sales data collection (Gittelsohn *et al.*, 2013).

12 Transforming local geographies to improve health

Louise Potvin

This book is about population health intervention research (PHIR) and space. More precisely, it attempts to make a case for the importance of geographers' involvement in this nascent field of scientific inquiry and, conversely, to convince population health intervention researchers to pay attention to geography as a constitutive discipline for this field of applied research. As a researcher whose work has been instrumental in the early development of PHIR, I certainly welcome the original contributions of this book to population health intervention research, a field that I have recently qualified as a "fundamental science for NCD prevention" (Potvin, 2013). Thus this book participates in shifting the research effort from identifying the causes of ill-health to developing the scientific basis for solving human health problems.

As a "science of solution," PHIR is about developing cumulative and transferable knowledge concerning population health interventions; how they are developed, implemented, sustained, and scaled up, as well as understanding how such interventions lead to transformative population health impacts (Hawe and Potvin, 2009). But how exactly are population health interventions defined and how do they differ from other health interventions? Population health interventions are generally understood as planned, coordinated actions in the form of policy or programs. Their planning and implementation involves a variety of stakeholders (Potvin *et al.*, 2005) in attempts to change the distribution of health risks in a whole population (Frieden, 2010), wherein a population is meant to be much more than the sum of its individuals (Krieger, 2010). Therefore intervening in the population's health is not merely a question of changing individual risk and behavior, one individual at a time. It is a system wide effort to transform the conditions in which risks occur in order to modify its distribution (Hawe, 2015). The fundamental unit for such interventions is a population. This, in turn, raises the question of what is a population.

In her extensive and well-documented essay on the meaning of population for population health, Krieger (2012b) emphasises the statistical origin of the term. Even as a statistical notion, a population should be understood more as a pool of individuals rather than a sum of individuals. That is, the total number of entities comprised in a given pool is itself an estimate defined by a set of explicit parameters instead of a factual reality. This stems from the fact that a population is dynamic and composite: "Populations are dynamic beings constituted by intrinsic relationships both among their members and with other populations that together produce their existence and make meaningful causal inference possible" (Krieger, 2012b: 363). Populations are thus a unique set of individuals, their relationships, and the structuring circumstances in which they evolve. In her eco-social theory of disease distribution, Krieger (2011) principally illustrates how the interactions between biological processes and those structuring forces shape the distribution of risk and ill-health in a population.

In his seminal 1977–8 lectures at the Collège de France, Michel Foucault (2004) explores another characteristic which is critical for human populations: populations are relational entities and objects of governmentality. By this Foucault means that in order to survive, human beings need to maintain social bounds with other human beings as well as exchange goods or services. These relationships and exchanges among humans are also inherent characteristics for defining a population. In turn, these relationships and exchanges among entities, holding varying values, interests, and projects, create a resource imbalance. This leads to the need for system coordination and regulation such that each entity can expect fair treatment and opportunities to pursue its own projects, achieve its interests, and fulfill its values. This is now known as governance. So, in addition to their demography, human populations are characterized by a sociology that describes their interrelationships by an economy comprised of their exchanges, and by a governance that describes how decisions regarding how to live collectively are made, by whom, and according to which rules.

Finally, human populations are situated in time and space. They have a history and a geography. These six dimensions (i.e. demography, sociology, economy, governance, history, and geography) form a set of interrelated dimensions that are critical for understanding human populations' functioning and dynamic as well as for attempting to improve their health through population health interventions. The collection of essays found in this book is a plea for geography, as a discipline, to play a central role in the development of population health intervention research.

Three ideas that shape the role for geography in population health intervention research

Health, as proposed by the Ottawa Charter for Health Promotion (World Health Organization, 1986), is produced in everyday life through the transformation of resources available in people's environments. Housing, food, clean air and water, positive human relationships, safety, and fulfilling jobs are among those resources locally available (or not) that have been associated with population health, and health inequalities (Macintyre and Ellaway, 2000). Although the relationship between health and such locally available resources appears to be robust, it is mediated by people's capacity to access and transform those resources into health (Bernard *at al.*, 2007). Health is a function of the quality, quantity, and accessibility of locally available resources and individuals' capacities (Abel and Frohlich,

2012). Thus, interventions in the form of programs or policy that seek to improve population health should aim to increase access to resources of sufficient quality and quantity through channels that are adequate and appropriate given people's capacities. Intervening to improve population health is intervening where people live, play, love, and work – in other words in their geography.

In his 2010 essay "Seeking Spatial Justice", the geographer Edward Soja (2010) argues that justice has a spatial dimension and that the equitable distribution of resources, services, goods, and all that is needed to lead a "good life" is of concern to geographers. The critical role of geography in relation to spatial justice and spatial consciousness is anchored in three fundamental ideas:

1. Geographies are not merely external containers of human activities ("given and immutable") for which only scale matters.
2. Geographies are not just the background or context that provides meaning for action; they are both the product and the conditions of human activity *in place*.
3. Geographies are amenable to changes through deliberate and planned social actions that can improve conditions for a better life.

> Moving closer toward a strategic spatial consciousness [. . .], it becomes evident and challenging that these socially produced geographies, because they are created by human actions can be changed and transformed through human agency. Human geographies are not merely external containers, given and immutable. Their changeability is crucial, for it makes our geographies the targets for social and political action seeking justice and democratic human rights by increasing their positive and/or decreasing their negative effects on our lives and livelihood.
>
> (Soja, 2010: 104)

With the understanding that health is also a human right and a prerequisite for a good life, and also that health, like social justice, is the product of the transformation of locally available resources, the collection of essays collated for this book contributes to transposing Soja's ideas to the health field. The introduction of geography, with its rich scientific tradition and conceptual apparatus, through this book, enriches the field of population health by enlightening the current debate among population health intervention researchers regarding the nature and role of context for population health interventions (Shoveller *et al.*, 2015).

Geographies are not merely external containers of human activities

The omnipresence of maps in our lives together with the ways in which public administrations operate the planning and provision of services according to administrative catchment areas have greatly contributed to the pervasive, and mostly false, impression that space is a container in which individuals can be counted, accounted for, and serviced. Several criticisms have emerged concerning the use of administratively defined territories in the planning and provisions of services (Potvin and Lamarre, 2009). First, these territories are mostly arbitrarily defined and seldom correspond to the "lived space" of individuals. Second, even using small scales, the individuals who constitute such territories are far from being homogeneous. Third and most importantly, individuals are mobile and their movements in space that connect the various places of their lives appear to be as important as the place in which their activity is located. Clearly space is an important dimension of the production of health and ill health as demonstrated by the robust correlations observed between spatial characteristics and health. However, this relationship appears to be complex and intervening "in space" to promote population health is not just a question of engineering physical environments.

This idea of the multi-dimensional nature of space is well captured in the chapter by Lewis (this volume) that discusses HIV prevention interventions for migrant populations. As shown in Lewis's chapter, population health interventions that focus on migrant populations provide an interesting example of the complex and evolving nature of space. By definition, migrants are a spatially mobile population. Further, through their culture, migrants carry with them a part of their original space within their new environment. Those various spaces interact and eventually clash, making the context much more difficult to interpret. To effectively intervene among migrant populations, in addition to the actual locations in which resources and services are to be made accessible, one needs also to take into account the meaning of those resources for migrant populations and the fact that these meanings can be quite different from those associated with the same resources for native populations.

Geographies are the product and conditions of human activity.

As a corollary to the proposition that space is more than a container for individuals' activities, the concept of relational space proposes that places and geographies are both products and conditions of human activity (Cummins *et al.*, 2007). Indeed, to the extent that people carry their space with them through their movement in space, the composition of shared geographies is constantly produced, transformed, and reproduced through human activities; space is dynamic and inseparable from the time dimension. Intervening to increase population health consists of modifying the characteristics of context (i.e. social and/or physical context) or the capacities of individuals in the population that influence the distribution of risk. Yet, at the same time, interventions are constrained by those same physical and social environments. This is precisely what is shown in Dicken's chapter in this book through the analysis of the interplay between exposure and vulnerability in the case of the risk of dengue, a communicable disease transmitted through an insect vector. Over and above climate conditions that determine the presence of a population of *Aedes*, the insect vector for dengue, this chapter demonstrates that human activity, in the form of land use and population density, influences the population's vulnerability to dengue. A truly socio-ecological view of human action not only suggests the important influence of the environment in constraining and enabling human activity, it also suggests that various species interact and influence each other's trajectory and in so doing transform their shared environment.

In Chapter 2 of this volume, Riley *et al.* show how partnerships of various forms shape the implementation of interventions in context. Indeed, the design and implementation of partnership-based interventions raise interesting questions on how to differentiate context and interventions. Interventions that are implemented and managed by a single organisation (whether or not the organisation is associated with research) leave other local organisations "in the context." If these other organisations interact with the intervention, they then eventually become beneficiaries or partners, depending on their level of investment and on whether they stay on the receiving or giving end of the intervention. This is also true for individuals and ultimately for all other human and non-human entities that compose the intervention's environment or context (Bisset *et al.*, 2009). As a consequence, the distinction between context and intervention is often quite arbitrary and depends mainly on one's conception of what interventions are made of (Potvin *et al.*, 2012). To the extent that interventions are conceived as open and adaptive systems which are, by definition, in constant interaction with their context, then context is, as geography tells us, a source of constraints and opportunities that constantly shape the intervention. So, in the same way that population interventions are attempts to (re)shape the geographies in which populations evolve, these interventions are in turn, conditioned by their implementation context.

Geographies are amenable to change through deliberate and planned social actions

Human action or agency (Poland *et al.*, 2008) is one of the forces that shape and transform natural and social environments. The assumption underlying population health intervention is that this force can be channelled and organised to pursue and achieve valued goals and objectives. Furthermore, in the case of PHIR, the main working hypothesis is that there exist patterns and regularity in the ways in which agency and context co-occur that allow the constitution of a body of knowledge that enables actors to anticipate the results of their action as a function of their action, in context. This idea of the possibility of creating a body of knowledge based on patterns identifiable across a number of interventions operating in a variety of contexts constitutes a major distinction between the field of evaluation and that of PHIR. Even if both fields of applied research share a broad range of methods, in evaluation the emphasis is often on the value judgement of the intervention that stems from its study using scientific methods. Conversely, in PHIR, the emphasis on the body of knowledge that single studies contribute to, constitute, and eventually form a "science of solution" (Potvin *et al.*, 2013).

Chapters 7, 10, and 11 in this book provide examples of how geography helps "contextualize" the problem/solution conception that underlies a population health intervention. The case of the distribution of bednets, treated or untreated, to reduce malaria transmission is a telling example of how the interaction of agency and context can be anticipated and put to work to increase the potential impact of an intervention (Janko and Emch, this volume). They demonstrate how impregnated bednets in malaria-ridden regions can have a double effect: on the one hand, it is a protective measure that lowers the risk of being bitten and of developing malaria for individuals who use them and, on the other hand, it changes the shared geography of mosquito and human populations in a way that preventing the spread of the virus in the human populations diminishes its prevalence in the mosquito population. In another example from this volume, Vine shows how the use of a geographical lens can inform the study of the implementation process of a school nutrition policy. She concludes that in the case of a school policy, the exploration of this setting with a geographical lens assists in identifying targets for interventions and areas where impacts can be anticipated and studied in the planning and conduct of evaluation. Finally, using various geographical concepts as metaphors to describe the availability and accessibility of food in given environments, Minaker *et al.* (this volume) show how taking into account those geographical dimensions can shape the planning and implementation of interventions in food retail outlets.

What's next?

I see this book as a first systematic attempt by health geographers to claim a role in the development of the growing field of population health intervention research. To the extent that population health interventions are operated by people in contexts to transform the conditions that shape the distribution of risk within a given population and to the extent that context is conceived as geographically situated, then the case should be clear: there is a role for geographers to play in the field of population health intervention. By applying the rich and diverse knowledge that geographers have accumulated about conceptually close notions such as space, place, environment, and so many others, they are key to help actors in population health interventions on both sides of practice and research, to strengthen our understanding of what context is made of, how it relates to interventions, and how to take it into account when developing and studying population health interventions. The time has passed when good health science was to control for, and eliminate, the role of context in our understanding of health phenomena. Understanding the intertwining of intervention and context is now essential for the development of population health programs and policy.

To illustrate the critical role of geography in this endeavour, I believe that we can paraphrase Soja's words:

> Moving closer toward a comprehensive approach to population health intervention [. . .], it becomes evident and challenging that socially produced contexts, because they are created by human actions, have to be changed and transformed through interventions. Contexts are not merely external

containers, given and immutable. Their changeability is crucial, for it makes our risk conducive contexts the targets for social and political action seeking justice and equity by increasing their positive and/or decreasing their negative effects on our health.

References

Abel, T. and Frohlich, K. L. (2012) "Capitals and capabilities: linking structure and agency to reduce health inequalities," *Social Science and Medicine*, 74(2): 236–44.

Acevedo-Garcia, D., Lochner, K. A., Osypuk, T. L., and Subramanian, S. V. (2003) "Future directions in residential segregation and health research: a multilevel approach," *American Journal of Public Health*, 93(2): 215–21.

Adam, B. D., Betancourt, G., and Serrano-Sanchez, A. (2011) "Development of an HIV prevention and life skills program for Spanish-speaking gay and bisexual newcomers to Canada," *Canadian Journal of Human Sexuality*, 20(1–2): 11–17.

Adami, H. O., Day, N. E., Trichopoulos, N. E., and Willett, W. C. (2001) "Primary and secondary prevention in the reduction of cancer morbidity and mortality," *European Journal of Cancer*, 37: S118–S127.

Adger, W. (2006) "Vulnerability," *Global Environmental Change*, 16(3): 268–81.

Adger, W. and Kelly, P. (1999) "Social vulnerability to climate change and the architecture of entitlements," *Mitigation and Adaptation Strategies for Global Change*, 4(3–4): 253–66.

Adger, W., Eakin, H., and Winkels, A. (2009) "Nested and teleconnected vulnerabilities to environmental change," *Frontiers in Ecology and the Environment*, 7(3): 150–7.

Afrane, Y. A., Little, T. J., Lawson, B. W., Githeko, A. K., and Yan, G. (2008) "Deforestation and vectorial capacity of Anopheles gambiae Giles mosquitoes in malaria transmission, Kenya," *Emerging Infectious Diseases*, 14(10): 1533–8.

Afrane, Y. A., Zhou, G., Lawson, B. W., Githeko, A. K., and Yan, G. (2006) "Effects of microclimatic changes caused by deforestation on the survivorship and reproductive fitness of Anopheles gambiae in western Kenya highlands," *American Journal of Tropical Medicine and Hygiene*, 74(5): 772–8.

Agriculture and Agri-Food Canada (2010) "Fresh foods: a growing opportunity in the North American convenience store sector" [online]. Available at: <http://www.ats-sea. agr.gc.ca/amr/5307-eng.htm> (accessed 27 August 2013).

Alcock, I., White, M. P., Wheeler, B. W., Fleming, L. E., and Depledge, M. H. (2013) "Longitudinal effects on mental health of moving to greener and less green urban areas," *Environmental Science and Technology*, 48(2): 1247–55.

Ali, M., Emch, M., von Seidlein, L., Yunus, M., Sack, D. A., *et al.* (2005) "Herd immunity conferred by killed oral cholera vaccines in Bangladesh: a reanalysis," *The Lancet*, 366(9479): 44–9.

Ali, M., Emch, M., Yunus, M., Sack, D., Lopez, A. L., Holmgren, J., *et al.* (2008) "Vaccine protection of Bangladeshi infants and young children against cholera:

implications for vaccine deployment and person-to-person transmission," *Pediatric Infectious Disease Journal*, 27(1): 33–7.

Alvaro, C., Jackson, L. A., Kirk, S., McHugh, T. L., Hughes, J., Chircop, A., and Lyons, R. F. (2011) "Moving Canadian governmental policies beyond a focus on individual lifestyle: some insights from complexity and critical theories," *Health Promotion International*, 26(1): 91–9.

Anderson, P. M. and Butcher, K. E. (2006) "Childhood obesity: trends and potential causes," *Future Child*, 16: 19–45.

Andrews, G. J. and Moon, G. (2005) "Space, place, and the evidence base: Part I – An introduction to health geography," *Worldviews on Evidence-Based Nursing*, 2(2): 55–62.

Anis, A. H., Zhang, W., Bansback, N., Guh, D. P., Amarsi, Z., and Birmingham, C. L. (2010) "Obesity and overweight in Canada: an updated cost-of-illness study," *Obesity Reviews*, 11(1): 31–40.

Anon. (2015) "Guiding Stars" [online]. Available at: <http://guidingstars.ca> (accessed 2 March 2015).

Apparicio, P., Cloutier, M. S., and Shearmur, R. (2007) "The case of Montreal's missing food deserts: evaluation of accessibility to food supermarkets," *International Journal of Health Geographics*, 6: 4.

Arnold, M., Hsu, L., Pipkin, S., McFarland, W., and Rutherford, G. W. (2009) "Race, place and AIDS: the role of socioeconomic context on racial disparities in treatment and survival in San Francisco," *Social Science and Medicine*, 69(1): 121–8.

Artiga, S., Stephens, J., and Damico, A. (2015) "The impact of the coverage gap in states not expanding Medicaid by race and ethnicity" [online]. Available at: <http://kff.org/disparities-policy/issue-brief/the-impact-of-the-coverage-gap-in-states-not-expanding-medicaid-by-race-and-ethnicity/> (accessed 25 May 2015).

Asanin Dean, J. and Elliott, S. J. (2012) "Prioritizing obesity in the city," *Journal of Urban Health*, 89(1): 196–213.

Association pour la Santé Publique du Québec (2011) "The school zone and nutrition: courses of action for the municipal sector" [pdf]. Available at: <http://www.aspq.org/documents/file/guide-zonage-version-finale-anglaise.pdf> (accessed 27 November 2015).

Ayanian, J., Weissman, J. S., Schneider, E. C., Ginsburg, J. A., and Zaslavsky, A. M. (2000) "Unmet health needs of uninsured adults in the United States," *Journal of American Medical Health Association*, 284(16): 2061–9.

Bailey, A. (2009) "Population geography: lifecourse matters," *Progress in Human Geography*, 33(3): 407–18.

Bailey, D. and Koney, K. M. (2000) *Strategic Alliances Among Health and Human Services Organizations*. Thousand Oaks, CA: Sage.

Batterman, S., Eisenberg, J., Hardin, R., Kruk, M., Lemos, M., Michalak, A., *et al.* (2009) "Sustainable control of water-related infectious diseases: a review and proposal for interdisciplinary health-based systems research," *Environmental Health Perspectives*, 117(7): 1023–32.

Baum, F. and Palmer, C. (2002) "'Opportunity structures': urban landscape, social capital and health promotion in Australia," *Health Promotion International*, 17(4): 351–61.

Baxter, J. and Eyles, J. (1997) "Evaluating qualitative research in social geography: establishing 'rigour' in interview analysis," *Transactions of the Institute of British Geographers*, 22(4): 505–25.

Berbés-Blázquez, M., Oestreicher, J., Mertens, F., and Saint-Charles, J. (2014) "Ecohealth and resilience thinking: a dialog from experiences in research and practice," *Ecology and Society*, 19(2): 24.

Bernard, P., Charafeddine, R., Frohlich, K. L., Daniel, M., Kestens, Y., and Potvin, L. (2007) "Health inequalities and place: a theoretical conception of neighbourhood," *Social Science and Medicine*, 65: 1839–52.

Best, A., Moor, G., Holmes, B., Clark, P. I., Brice, T., Lieschow, S., *et al.* (2003) "Health promotion dissemination and system thinking: towards an integrative model," *American Journal of Health Behavior*, 27 (Suppl): S206–S216.

Best, A., Stokols, D., Green, L. W., Leischow, S., Holmes, B., and Buchholz, K. (2003) "An integrative framework for community partnering to translate theory into effective health promotion strategy," *American Journal of Health Promotion*, 18(2): 168–76.

Beyer, K. M., Kaltenbach, A., Szabo, A., Bogar, S., Nieto, F. J. and Malecki, K. M. (2014) "Exposure to neighborhood green space and mental health: evidence from the survey of the health of Wisconsin," *International Journal of Environmental Research and Public Health*, 11(3): 3453–72.

Bianchi, F. T., Reisen, C. A., Zea, M. C., Poppen, P. J., and Shedlin, M. J. (2007) "The sexual experiences of Latino men who have sex with men who migrated to a gay epicentre in the USA," *Culture, Health and Sexuality*, 9(5): 505–18.

Binka, F. N., Kubaje, A., Adjuik, M., Williams, L. A., Lengeler, C., Maude, G. H., *et al.* (1996) "Impact of permethrin impregnated bednets on child mortality in Kassena-Nankana district, Ghana: a randomized controlled trial," *Tropical Medicine and International Health*, 1(2): 147–54.

Birkmann, J., Cardona, O., Carreño, M., Barbat, A., Pelling, M., Schneiderbauer, S., *et al.* (2013) "Framing vulnerability, risk and societal responses: the MOVE framework," *Natural Hazards*, 67(2): 193–211.

Bisset, S., Daniel, M., and Potvin, L. (2009) "Exploring the intervention context interface: a case from a school based nutrition intervention,"*American Journal of Evaluation*, 30: 554–71.

Black, J. (2014) "Local food environments outside of the United States – a look to the north: examining food environments in Canada," in K. B. Morland (ed.), *Local Food Environments: Food Access in America*. Boca Raton, FL: CRC Press, pp. 231–61.

Black, J. L. and Day, M. (2012) "Availability of limited service food outlets surrounding schools in British Columbia," *Canadian Journal of Public Health*, 103(4): 255–9.

Black, J. L. and Macinko, J. (2008) "Neighborhoods and obesity," *Nutrition Reviews*, 66(1): 2–20.

Black, J. L., Carpiano, R. M., Fleming, S., and Lauster, N. (2011) "Exploring the distribution of food stores in British Columbia: associations with neighbourhood sociodemographic factors and urban form," *Health and Place*, 17(4): 961–70.

Boeing, H., Bechthold, A., Bub, A., Ellinger, S., Haller, D., Kroke, A., *et al.* (2012) "Critical review: vegetables and fruit in the prevention of chronic diseases," *European Journal of Nutrition*, 51(6): 637–63.

Bonell, C., Parry, W., Wells, H., Jamal, F., Fletcher, A., Harden, A., *et al.* (2013) "The effects of the school environment on student health: a systematic review of multi-level studies," *Health and Place*, 21: 180–91.

Borrell, C., Rodríguez-Sanz, M., Pasarín, M. I., Brugal, M. T., García-de-Olalla, P., Marí-Dell'Olmo, M., *et al.* (2006) "AIDS mortality before and after the introduction of highly active antiretroviral therapy: does it vary with socioeconomic group in a country with a National Health System?" *European Journal of Public Health*, 16(6): 601–8.

Bowen, A., Horvath, K., and Williams, M. (2006) "Randomized control trial of an Internet delivered HIV knowledge intervention with MSM," *Health Education and Research*, 22(1): 120–7.

Bowen, K., Ebi, K., Friel, S., and McMichael, A. (2013) "A multi-layered governance framework for incorporating social science insights into adapting to the health impacts of climate change," *Global Health Action*, 6 [online]. Available at: <http://www. globalhealthaction.net/index.php/gha/article/view/21820> (accessed 9 April 2016).

Boydell, L. R. and Rugkasa, J. (2007) "Benefits of working in partnership: a model," *Critical Public Health*, 17(3): 203–14.

Boydell, L. R., Hoggett, P., Rugkasa, J., and Cummins, A. M (2008) "Intersectoral partnerships, the knowledge economy and intangible assets," *Policy and Politics*, 36(2): 209–24.

Branas, C. C. (2013) "Safe and healthy places are made not born" [online]. Available at: <http://penniur.upenn.edu/publications/sa> (accessed 13 September 2013).

Branas, C. C. and Macdonald, J. M. (2014) "A simple strategy to transform health, all over the place," *Journal of Public Health Management and Practice*, 20(2): 157–9.

Branas, C. C., Cheney, R. A., MacDonald, J. M., Tam, V. W., Jackson, T. D., and Ten Have, T. R. (2011) "A difference-in-differences analysis of health, safety, and greening vacant urban space," *American Journal of Epidemiology*, 174(11): 1296–306.

Bronfenbrenner, U. (1979) *The Ecology of Human Development: Experiments by Nature and Design*. Cambridge, MA: Harvard University Press.

Brown, C. and Wilby, R. (2012) "An alternate approach to assessing climate risks," *Eos, Transactions American Geophysical Union*, 93(41): 401.

Brown, E. R., Ojeda, V. D., Wyn, R., and Levan, R. (2000) *Racial and Ethnic Disparities in Access to Health Insurance and Health Care*. Los Angeles: UCLA Center for Health Policy Research and the Henry J. Kaiser Family Foundation.

Brown, I. (2014) City of Milwaukee, personal communication.

Brown, T., McLafferty, S., and Moon, G. (2009) "Introduction to health and medical geography," in T. Brown, S. McLafferty, and G. Moon (eds), *A Companion to Health and Medical Geography*. Oxford: Wiley-Blackwell, ch. 1.

Browning, C. R. and Cagney, K. A. (2002) "Neighborhood structural disadvantage, collective efficacy, and self-rated physical health in an urban setting," *Journal of Health and Social Behavior*, 43(4): 383–99.

Brulle, R. J. and Pellow, D. N. (2006) "Environmental justice: human health and environmental inequalities," *Annual Review of Public Health*, 27: 103–24.

Bucchianeri, G. W. and Wachter, S. M. (2007) "What is a tree worth? Green-city strategies and housing prices" (with Susan Wachter), *Real Estate Economics*, 36(2): 213–39.

Bucchianeri, G. W., Gillen, K. C. and Wachter, S. M. (2012) "Valuing the Conversion of Urban Greenspace" [pdf]. Available at: <http://phsonline.org/media/resources/ Bucchianeri_Gillen_Wachter_Valuing_Conversion_Urban_Greenspace_Final_Draft_ KG_changesacceptes.pdf> (accessed 27 November 2015).

Bull, S., Pratte, K., Whitesell, N., Rietmeijer, C., and McFarlane, M. (2009) "Effects of an Internet-based intervention for HIV prevention: the Youthnet trials," *AIDS and Behavior*, 13(3): 474–87.

Burkholder, S. (2012) "The new ecology of vacancy: rethinking land use in shrinking cities," *Sustainability*, 4(6): 1154–72.

Bursik, R. J. (1988) "Social disorganization and theories of crime and delinquency: problems and prospects," *Criminology*, 26(4): 519–52.

Burton, I., Kates, R., and White, G. (1978) *The Environment as Hazard*. New York: Oxford University Press.

Business Dictionary (2014) "Organizational Learning" [online]. Available at: <http://www.businessdictionary.com/definition/organizational-learning.html> (accessed 1 August 2014).

Cameron, R., Manske, S., Brown, K., Jolin, M., Murnaghan, D. and Lovato, C. (2007) "Integrating public health policy, practice, evaluation, surveillance, and research using local data collection and feedback systems: a prototype focused on schools," *American Journal of Public Health*, 97: 648–54.

Canadian Institute for Health Information (2014) "Population Health Intervention Research for Canada (PHIRIC)" [online]. Available at: <http://www.cihi.ca/CIHI-ext-portal/internet/en/document/factors+influencing+health/environmental/cphi_phiric> (accessed 18 June 2015).

Canadian Institutes of Health Research (2012) "Population Health Intervention Research Initiative for Canada" [online]. Available at: <http://www.cihr-irsc.gc.ca/e/38731.html> (accessed 18 June 2015).

Carballo-Diéguez, A., Dolezal, C., Leu, C.-S., Nieves, L., Diaz, F., Decena, C., *et al.* (2005) "A randomized controlled trial to test an HIV-prevention intervention for Latino gay and bisexual men: lessons learned," *AIDS Care*, 17(3): 314–28.

Carrillo, H. (2004) "Sexual migration, cross-cultural sexual encounters, and sexual health," *Sexuality Research and Social Policy*, 1(3): 58–70.

Caspi, C. E., Sorensen, G., Subramanian, S. V., and Kawachi, I. (2012) "The local food environment and diet: a systematic review," *Health and Place*, 18: 1172–87.

Castilla, J., Sobrino, P., de la Fuente, L., Noguer, I., Guerra, L., and Parras, F. (2002) "Late diagnosis of HIV infection in the era of highly active antiretroviral therapy: consequences for AIDS incidence," *AIDS*, 16(14): 1945–51.

Center for Health Law and Policy Innovation of Harvard Law School and the Treatment Access Expansion Project (2013) *State Health Reform Impact Modeling Project: Texas* [pdf]. Available at: <http://www.hivhealthreform.org/wp-content/uploads/2013/03/Texas-Modeling-Report-Final.pdf> (accessed 3 April 2016).

Centers for Disease Control and Prevention (2012) "HIV Diagnosis" [online]. Available at: <http://gis.cdc.gov/GRASP/NCHHSTPAtlas/main.html> (accessed 3 March 2013).

Centers for Disease Control and Prevention (2013a) "HIV at a Glance" [pdf]. Available at: <http://www.cdc.gov/hiv/resources/factsheets/PDF/stats_basics_factsheet.pdf> (accessed 10 February 2013).

Centers for Disease Control and Prevention (2013b) "Diagnosed HIV infection among adults and adolescents in metropolitan statistical areas – United States and Puerto Rico, 2011," *HIV Surveillance Supplemental Report*, 18(8).

Centers for Disease Control and Prevention (2014) "NCHHSTP Atlas: Persons Living with Diagnosed HIV 2011" [online]. Available at: <http://gis.cdc.gov/GRASP/NCHHSTPAtlas/main.html> (accessed 22 December 2014).

Centers for Disease Control and Prevention (2016) *Minority Health* [online]. Available at: <http://www.cdc.gov/minorityhealth/> (accessed 9 April 2016).

Chan, W. C. and Leatherdale, S. T. (2011) "Tobacco retailer density surrounding schools and youth smoking behaviour: a multi-level analysis," *Tobacco Induced Diseases*, 9(9).

Chen, H. Y., Baumgardner, D. J., Galvao, L. W., Rice, J. P., Swain, G. R., and Cisler, R. A. (2011) *Milwaukee Health Report 2011: Health Disparities in Milwaukee by Socioeconomic Status*. Milwaukee, WI: Center for Urban Population Health.

Choi, K. H., Lew, S., Vittinghoff, E., Catania, J. A., Barrett, D. C., and Coates, T. J. (1996) "The efficacy of brief group counseling in HIV risk reduction among homosexual Asian and Pacific Islander men," *AIDS*, 10(1): 81–8.

Choi, W. S., Novotny, T. E., Davis, R. M., and Epstein, J. (1991) "State tobacco prevention and control activities: results of the 1989–1990 Association of State and Territorial Health Officials (ASTHO) Survey Final Report," *Morbidity and Mortality Weekly Report: Recommendations and Reports*, pp. i–40.

City of Milwaukee (2013) *Vacant Lot Handbook: A Guide to Reusing, Reinventing and Adding Value to Milwaukee's City-owned Vacant Lots*.

City of Milwaukee (2014a) "Parcel Database" [online]. Available at: <http://city.milwaukee. gov/DownloadMapData3497.htm#.VJM8wclNqQi> (accessed 20 October 2014).

City of Milwaukee (2014b) "Master Property Record Database (MPROP)" [online]. Available at: <http://city.milwaukee.gov/DownloadTabularData3496.htm#. VJM1J8INqQg> (accessed 24 October 2014).

City of Milwaukee Department of City Development (2013) *City of Milwaukee: Vacant Lot Pattern Book*.

Clarke, S. E., Bøgh, C., Brown, R. C., Pinder, M., Walraven, G. E. L., and Lindsay, S. W. (2001) "Do untreated bednets protect against malaria?" *Transactions of the Royal Society of Tropical Medicine and Hygiene*, 95(5): 457–62.

Cleveland Neighborhood Progress (2014) "Re-imagining Cleveland: Ideas to Action Resource Book" [pdf]. Available at: <http://www.npi-cle.org/files/2012/07/ IdeastoActionResourceBook.pdf> (accessed 15 May 2014).

Cohen, D. A., Mason, K., Bedimo, A., Scribner, R., Basolo, V., and Farley, T. A. (2003) "Neighborhood physical conditions and health," *American Journal of Public Health*, 93(3): 467–71.

Colditz, G. A., DeJong, W., Hunter, D., Trichopoulos, D., and Willett, W. (1996) "Harvard report on cancer prevention. Volume 1: Causes of human cancer," *Cancer Causes and Control*, 7: S3–S58.

Community Intervention Trial for Smoking Cessation (COMMIT) (1995a) "Community Intervention Trial for Smoking Cessation (COMMIT): I. Cohort Results from a four-year community intervention," *American Journal of Public Health*, 85(2): 183–92.

Community Intervention Trial for Smoking Cessation (COMMIT) (1995b) "Community Intervention Trial for Smoking Cessation (COMMIT): II. Changes in adult cigarette smoking prevalence," *American Journal of Public Health*, 85(2): 193–200.

Confalonieri, U., Lima, A., Brito, I., and Quintão, A. (2013) "Social, environmental and health vulnerability to climate change in the Brazilian Northeastern Region," *Climatic Change*, 17(1): 123–37.

Conference of Principal Investigators of Heart Health (2001) *Canadian Heart Health Initiative: Process Evaluation of the Demonstration Phase*. Ottawa: Health Canada.

Conner, R. F., Takahashi, L., Ortiz, E., Archuleta, E., Muniz, J., and Rodriguez, J. (2005) "The SOLAAR HIV Prevention Program for gay and bisexual Latino men: using social marketing to build capacity for service provision and evaluation," *AIDS Education and Prevention*, 17(4): 361–74.

Cook, T. D. and Campbell, D. T. (1979) *Quasi-experimentation: Design and Analysis Issues for Field Settings*. Boston: Houghton Mifflin.

Cope, M. and Elwood, S. (2009) *Qualitative GIS: A Mixed Methods Approach*. Thousand Oaks, CA: Sage.

Corbett, K., Thompson, B., White, N., and Taylor, S. M. (1990) "Process evaluation in the Community Intervention Trial for Smoking Cessation (COMMIT)," *International Quarterly of Community Health Education*, 11(3): 291–309.

Corvalán, C., McMichael, A., and Hales, S. (2005) *Ecosystems and Human Well-being*. Geneva: World Health Organization.

Costello, A., Abbas, M., Allen, A., Ball, S., Bell, S., Bellamy, R., *et al.* (2009) "Managing the health effects of climate change," *The Lancet*, 373(9676): 1693–733.

Costello, M. J. E., Leatherdale, S. T., Ahmed, R., Church, D. L., and Cunningham, J. A. (2012) "Co-morbid substance use behaviors among youth: any impact of school environment?" *Global Health Promotion*, 19: 51–8.

Cote, M. and Nightingale, A. (2012) "Resilience thinking meets social theory: situating social change in socio-ecological systems (SES) research," *Progress in Human Geography*, 36(4): 475–89.

Covington, J. and Taylor, R. B. (1991) "Fear of crime in urban residential neighborhoods," *Sociological Quarterly*, 32(2): 231–49.

Craig, P., Dieppe, P., Macintyre, S., Michie, S., Nazareth, I., and Petticrew, M. (2008) "Developing and evaluating complex interventions: the new Medical Research Council guidance," *BMJ*, 337: a1655 [online]. Available at: <http://www.bmj.com/content/337/bmj.a1655> (accessed 16 July 2015).

Craig, P., Cooper, C., Gunnell, D., Haw, S., Lawson, K., Macintyre, S., *et al.* (2012) "Using natural experiments to evaluate population health interventions: new Medical Research Council guidance," *Journal of Epidemiology and Community Health*, 66(12): 1182–6.

Crepaz, N., Marshall, K. J., Aupont, L. W., Jacobs, E. D., Mizuno, Y., Kay, L. S., *et al.* (2009) "The efficacy of HIV/STI behavioral interventions for African American females in the United States: a meta-analysis," *American Journal of Public Health*, 99(11): 2069–78.

Crilly, T., Jashapara, A. and Ferlie, E. (2010) *Research Utilisation and Knowledge Mobilisation: A Scoping Review of the Literature*, Report for the National Institute for Health Research Service Delivery and Organisation Programme [pdf]. Available at: <http://www.netscc.ac.uk/hsdr/files/project/SDO_FR_08-1801-220_V01.pdf> (accessed 27 November 2015).

Crutzen, P. (2002) "Geology of mankind," *Nature*, 415, 3 January, p. 23.

Cummins, S. and Macintyre, S. (2006) "Food environments and obesity – neighbourhood or nation?' *International Journal of Epidemiology*, 35(1): 100–4.

Cummins, S., Curtis, S., Diez-Roux, A. V. and Macintyre, S. (2007) "Understanding and representing 'place' in health research: a relational approach," *Social Science and Medicine*, 65(9): 1825–38.

Cutter, S., Barnes, L., Berry, M., Burton, C., Evans, E., *et al.* (2008) "A place-based model for understanding community resilience to natural disasters," *Global Environmental Change*, 18(4): 598–606.

D'alessandro, U., Olaleye, B., Langerock, P., Aikins, M. K., Thomson, M. C., Cham, M. K., *et al.* (1995) "Mortality and morbidity from malaria in Gambian children after introduction of an impregnated bednet programme," *The Lancet*, 345(8948): 479–83.

D'alessandro, U., Olaleye, B. O., McGuire, W., Thomson, M. C., Langerock, P., Bennett, S., *et al.* (1995) "A comparison of the efficacy of insecticide-treated and untreated bed nets in preventing malaria in Gambian children," *Transactions of the Royal Society of Tropical Medicine and Hygiene*, 89(6): 596–8.

Daniel, M., Kestens, Y., and Paquet, C. (2009) "Demographic and urban form correlates of healthful and unhealthful food availability in Montreal, Canada," *Canadian Journal of Public Health/Revue Canadienne de Santé Publique*, 100(3): 189–93.

Dauchet, L., Amouyel, P., and Dallongeville, J. (2009) "Fruits, vegetables and coronary heart disease," *Nature Reviews: Cardiology*, 6(9): 599–608.

Davidson, J. and Milligan, C. (2004) "Embodying emotion, sensing space: introducing emotional geographies," *Social and Cultural Geography*, 5(4): 523–32.

de Sherbinin, A. (2014) *Spatial Climate Change Vulnerability Assessments: A Review of Data, Methods, and Issues*, African and Latin American Resilience to Climate Change Project Report. USAID.

Dear, M. and Wolch, J. (1987) *Landscapes of Despair: From Deinstitutionalization to Homelessness*. Princeton, NJ: Princeton University Press.

Department for International Development (DFID) (1999) *Sustainable Livelihoods Guidance Sheets*. London: DFID.

Detroit Future City (2014) *Vacant Land Treatment Program* [pdf]. Available at: <http://www.detroitfuturecity.com/wp-content/uploads/2014/03/dfc-vacant-land-treatment-20140228.pdf> (accessed 15 May 2014).

Diaz, R. M., Ayala, G., Bein, E., Henne, J., and Marin, B. V. (2001) "The impact of homophobia, poverty, and racism on the mental health of gay and bisexual Latino men: findings from 3 U.S. cities," *American Journal of Public Health*, 91: 927–32.

Dickin, S. and Schuster-Wallace, C. (2014) "Assessing changing vulnerability to dengue in northeastern Brazil using a water-associated disease index approach," *Global Environmental Change*, 29: 155–64.

Dickin, S., Schuster-Wallace, C., and Elliott, S. (2013) "Developing a vulnerability mapping methodology: applying the water-associated disease index to dengue in Malaysia," *PLoS ONE*, 8(5): e63584.

Diez-Roux, A. V. (1998) "Bringing context back into epidemiology: variables and fallacies in multilevel analysis," *American Journal of Public Health*, 88(2): 216–22.

DiStefano, C., Zhu, M,. and Mindrila, D. (2009) "Understanding and using factor scores: considerations for the applied researcher," *Practical Assessment, Research and Evaluation*, 14(20): 1–11.

DMTI Spatial Inc. (2014) "CanMap Route Logistics (CMRL) & Enhanced Points of Interest (EOPI)" [online]. Available at: <http://www.dmtispatial.com> (accessed 27 November 2015).

Dowling, B., Powell, M., and Glendinning, C. (2004) "Conceptualising successful partnerships," *Health and Social Care in the Community*, 12(4): 309–17.

Downs, S. M., Farmer, A., Quintanilha, M., Berry, T. R., Mager, D. R., Willows, N. D., et al. (2011) "Alberta Nutrition Guidelines for Children and Youth: awareness and use in schools," *Canadian Journal of Dietetic Practice and Research*, 72(3): 137–40.

Downs, S. M., Farmer, A., Quintanilha, M., Berry, T. R., Mager, D. R., Willows, N. D., et al. (2012) "From paper to practice: barriers to adopting nutrition guidelines in schools," *Journal of Nutrition Education and Behavior*, 44(2): 114–22.

Drouin, S., Hamelin, A. M., and Quellet, D. (2009) "Economic access to fruits and vegetables in the greater Quebec City: do disparities exist?" *Canadian Journal of Public Health*, 100(5): 361–4.

Drucker, E., Alcabes, P., Sckell, B., and Bosworth, W. (1994) "Childhood tuberculosis in the Bronx, New York," *The Lancet*, 343(8911): 1482–5.

Duff, C. (2011) "Networks, resources and agencies: on the character and production of enabling places," *Health and Place*, 17(1): 149–56.

Dunn, C. E. (2007) "Participatory GIS – a people's GIS?" *Progress in Human Geography*, 31(5): 616–37.

Dunn, J. R. (2014) "Evaluating place-based programmes for health improvement," *Journal of Epidemiology and Community Health*, 68(7): 591.

Eakin, H. and Luers, A. (2006) "Assessing the vulnerability of social-environmental systems," *Annual Review of Environment and Resources*, 31(1): 365–94.

Ebi, K., Kovats, R., and Menne, B. (2006) "An approach for assessing human health vulnerability and public health interventions to adapt to climate change," *Environmental Health Perspectives*, 114(12): 1930–4.

Edwards, N., Mill, J., and Kothari, A. R. (2004) "Multiple intervention research programs in community health," *Canadian Journal of Nursing Research*, 36(1): 40–54.

Egan, J. E., Frye, V., Kurtz, S. P., Latkin, C., Chen, M., Tobin, K., *et al.* (2011) "Migration, neighborhoods, and networks: approaches to understanding how urban environmental conditions affect syndemic adverse health outcomes among gay, bisexual and other men who have sex with men," *AIDS and Behavior*, 15(1): 35–50.

Egger, G. and Swinburn, B. (1997) "An 'ecological' approach to the obesity pandemic," *British Medical Journal*, 315(7106): 477–80.

Egger, M., May, M., Chêne, G., Phillips, A. N., Ledergerber, B., Dabis, F., *et al.* (2002) "Prognosis of HIV-1-infected patients starting highly active antiretroviral therapy: a collaborative analysis of prospective studies," *The Lancet*, 360(9327): 119–29.

Eisele, T., Macintyre, K., Eckert, E., Beier, J., and Killeen, G. (2000) "Evaluating malaria interventions in Africa: a review and assessment of recent research" [online]. Available at: < http://www.cpc.unc.edu/measure/resources/publications/wp-00-19> (accessed 27 November 2015).

Eisenberg, J., Cevallos, W., Ponce, K., Levy, K., Bates, S., Scott, J., *et al.* (2006) "Environmental change and infectious disease: how new roads affect the transmission of diarrheal pathogens in rural Ecuador," *Proceedings of the National Academy of Sciences*, 103(51): 19460–5.

Eisenberg, J., Desai, M., Levy, K., Bates, S., Liang, S., Naumoff, K., *et al.* (2007) "Environmental determinants of infectious disease: a framework for tracking causal links and guiding public health research," *Environmental Health Perspectives*, 115(8): 1216–23.

Elliott, S. J., O'Loughlin, J., Robinson, K., Eyles, J., Cameron, R., Harvey, D., *et al.* (2003) "Conceptualizing dissemination research and activity: the case of the Canadian Heart Health Initiative," *Health Education and Behavior*, 30(3): 267–82.

Emch, M., Ali, M., Park, J. K., Yunus, M., Sack, D. A., and Clemens, J. D. (2006) "Neighborhood-level ecological correlates of cholera vaccine protection," *International Journal of Epidemiology*, 35: 1044–50.

Emch, M., Ali, M., Acosta, C., Yunus, M., Sack, D. A., and Clemens, J. D. (2007) "Efficacy calculation in randomized trials: global or local measures?" *Health and Place*, 13(1): 238–48.

Engler-Stringer, R., Le, H., Gerrard, A., and Muhajarine, N. (2014) "The community and consumer food environment and children's diet: a systematic review," *BMC Public Health*, 14: 522.

Engler-Stringer, R., Shah, T., Bell, S., and Muhajarine, N. (2014) "Geographic access to healthy and unhealthy food sources for children in neighbourhoods and from elementary schools in a mid-sized Canadian city," *Spatial and Spatio-Temporal Epidemiology*, 11: 23–32.

Entman, R. M. (1993)" Framing: toward clarification of a fractured paradigm," *Journal of Communication*, 43(4): 51–8.

Environmental Protection Agency (2014) "EnviroAtlas" [online]. Available at: <http://enviroatlas.epa.gov/EnviroAtlas/>.

Eriksson, M. (2011) "Social capital and health – implications for health promotion," *Global Health Action*, 4: 5611.

Escaron, A. L., Meinen, A. M., Nitzke, S. A., and Martinez-Donate, A. P. (2013) "Supermarket and grocery store-based interventions to promote healthful food choices and eating practices: a systematic review," *Preventing Chronic Disease*, 10(4): 1–20.

ESRI (2014) "ArcGIS Collector" [online]. Available at: <http://doc.arcgis.com/en/collector/> (accessed 2 December 2014).

Evans, R. G. and Stoddart, G. L. (1994) "Producing health, consuming health care," *Social Science and Medicine*, 31(12): 1347–63.

Fekete, A. (2009) "Validation of a social vulnerability index in context to river-floods in Germany," *Natural Hazards and Earth System Science*, 9(2): 393–403.

Feng, J., Glass, T. A., Curriero, F. C., Stewart, W. F., and Schwartz, B. S. (2010) "The built environment and obesity: a systematic review of the epidemiologic evidence," *Health and Place*, 16(2): 175–90.

Fernández, M. (2006) "Pros and cons of a zoning diet: fighting obesity by limiting fast-food restaurants," *New York Times* [online], 24 September. Available at: <http://www.nytimes.com/2006/09/24/nyregion/24fast.html?_r=0> (accessed 27 November 2015).

Fernández, M. I., Jacobs, R. J., Warren, J. C., Sanchez, J., and Bowen, G. S. (2009) "Drug use and Hispanic men who have sex with men in South Florida: implication for intervention development," *AIDS Education and Prevention*, 21(Supplement B): 45–60.

Few, R. (2007) "Health and climatic hazards: framing social research on vulnerability, response and adaptation," *Global Environmental Change*, 17(2): 281–95.

Fielding, J. E. and Simon, P. A. (2011) "Food deserts or food swamps?" *Archives of Internal Medicine*, 171(13): 1171–2.

Finkelstein, D. M., Hill, E. L., and Whitaker, R. C. (2008) "School food environments and policies in US public schools," *Pediatrics*, 122(1): e251–e259.

Finley, R. (2014) "Survey Monkey" [online]. Available at: <https://www.surveymonkey.com/home/> (accessed 23 November 2014).

Fiscella, K. and Williams, D. R. (2004) "Health disparities based on socioeconomic inequities: implications for urban health care," *Academic Medicine*, 79(12): 1139–47.

Fisher Jr, E. B. (1995) "Editorial: the results of the COMMIT trial," *American Journal of Public Health*, 85(2): 159–60.

FoodShare (2013) "About us – FoodShare Toronto" [online]. Available at: <http://www.foodshare.net/about-us-2> (accessed 20 September 2013).

Ford, P. B. and Dzewaltowski, D. A. (2008) "Disparities in obesity prevalence due to variation in the retail food environment: three testable hypotheses," *Nutrition Reviews*, 66(4): 216–28.

Forsyth, A., Lytle, L. A., and Van Riper, D. (2010) "Finding food: issues and challenges in using geographic information systems to measure food access," *Journal of Transport and Land Use*, 3(1): 43–65.

Foster-Fishman, P. G., Nowell, B. and Yang, H. (2007) "Putting the system back into systems change: a framework for understanding and changing organizational and community systems," *American Journal of Community Psychology*, 39(3–4): 197–215.

Foucault, M. (2004) *Sécurité, territoire, population*. Cours au Collège de France, 1977–8. Paris: Collection Hautes études, EHESS, Gallimard, Seuil.

Frank, J. W. (1995) "Why 'population health?'," *Canadian Journal of Public Health*, 86(3): 162–4.

Frank, J. (2012) "Public health interventions to reduce inequalities: what do we know works?" *Canadian Journal of Public Health*, 103(Suppl. 1): S5–S7.

Frank, L. D. and Engelke, P. O. (2001) "The built environment and human activity patterns: exploring the impacts of urban form on public health," *Journal of Planning Literature*, 16(2): 202–18.

Frankish, C. J., Moulton, G. E., Quantz, D. Q., Carson, A. J., Casebeer, A. L., Eyles, J. D., *et al.* (2007) "Addressing non-medical determinants of health: a survey of Canada's health regions," *Canadian Journal of Public Health*, 98(1): 41–7.

Frankish, J. (2012) "Population health intervention research: advancing the field," *Canadian Journal of Public Health*, 103(Suppl. 1): S3–S4.

Frieden, T. R. (2010) "A framework for public health action: the health impact pyramid," *American Journal of Public Health*, 100(4): 590–5.

Fuller, D. and Potvin, L. (2012) "Context by treatment interactions as the primary object of study in cluster randomized controlled trials of population health interventions," *International Journal of Public Health*, 57(3): 633–6.

Fuller, D., Gauvin, L., Kestens, Y., Daniel, M., Fournier, M., Morency, P., *et al.* (2011a) "Use of a new public bicycle share program in Montreal, Canada," *American Journal of Preventive Medicine*, 41(1): 80–3.

Fuller, D., Sabiston, C., Karp, I., Barnett, T., and O'Loughlin, J. (2011b) "School sports opportunities influence physical activity in secondary school and beyond," *Journal of School Health*, 81(8): 449–54.

Fuller, D., Gauvin, L., Kestens, Y., Daniel, M., Fournier, M., Morency, P. *et al.* (2013a) "Impact evaluation of a public bicycle share program on cycling: a case example of BIXI in Montreal, Quebec," *American Journal of Public Health*, 103(3): e1–e8.

Fuller, D., Gauvin, L., Morency, P., Kestens, Y. and Drouin, L. (2013b) "The impact of implementing a public bicycle share program on the likelihood of collisions and near misses in Montreal, Canada," *Preventive Medicine*, 57(6): 920–4.

Fuller, D., Gauvin, L., Kestens, Y., Morency, P. and Drouin, L. (2013c) "The potential modal shift and health benefits of implementing a public bicycle share program in Montreal, Canada," *International Journal of Behavioral Nutrition and Physical Activity*, 10(1): 66.

Fuller, D., Gauvin, L., Dubé, A.-S., Winters, M., Tesche, K., Russo, E. T., *et al.* (2014) "Evaluating the impact of environmental interventions across 2 countries: the International Bikeshare Impacts on Cycling and Collisions Study (IBICCS) Study protocol," *BMC Public Health*, 14(1): 1103.

Fung, C., McIsaac, J. L. D., Kuhle, S., Kirk, S. F. L., and Veugelers, P. J. (2013) "The impact of a population-level school food and nutrition policy on dietary intake and body weights of Canadian children," *Preventive Medicine*, 57(6): 934–40.

Gajda, R. (2004) "Utilizing collaboration theory to evaluate strategic alliances," *American Journal of Evaluation*, 25(1): 65–77.

Galea, S. and Vlahov, D. (2005) "Urban health: evidence, challenges, and directions," *Public Health*, 26: 341–65.

Garriguet, D. (2004) *Overview of Canadians' Eating Habits* (Catalogue no. 82-620-MIE – No. 2). Ottawa: Statistics Canada.

Garriguet, D. (2007) "Canadians' eating habits," *Health Reports/Statistics Canada: Canadian Centre for Health Information / Rapports sur la Santé/Statistique Canada: Centre Canadien d'Information sur la Santé*, 18(2): 17–32.

Garriguet, D. (2009) *Diet Quality in Canada*. Ottawa: Statistics Canada.

Garvin, E., Cannuscio, C., and Branas, C. (2013) Greening vacant lots to reduce violent crime: a randomised controlled trial," *Injury Prevention: Journal of the International Society for Child and Adolescent Injury Prevention*, 19(3): 198–203.

Garvin, E., Branas, C., Keddem, S., Sellman, J., and Cannuscio, C. (2013) "More than just an eyesore: local insights and solutions on vacant land and urban health," *Journal of Urban Health*, 90(3): 412–26.

Gatrell, A. (2005) "Complexity theory and geographies of health: a critical assessment," *Social Science and Medicine*, 60(12): 2661–71.

Gatrell, A. C. and Elliott, S. J. (2009) *Geographies of Health: An Introduction*, 2nd edn. Malden, MA: Blackwell.

Genton, B., Hii, J., Al-Yaman, F., Paru, R., Beck, H. P., Ginny, M., *et al.* (1994) "The use of untreated bednets and malaria infection, morbidity and immunity," *Annals of Tropical Medicine and Parasitology*, 88(3): 263–70.

Geronimus, A. T. (2000) "To mitigate, resist, or undo: addressing structural influences on the health of urban populations," *American Journal of Public Health*, 90(6): 867–72.

Gesler, W. M. (1992) "Therapeutic landscapes: medical issues in light of the new cultural geography," *Social Science and Medicine*, 34(7): 735–46.

Gething, P. W., Patil, A. P., Smith, D. L., Guerra, C. A., Elyazar, I. R., Johnston, G. L., *et al.* (2011) "A new world malaria map: Plasmodium falciparum endemicity in 2010," *Malaria Journal*, 10(378): 1475–2875.

Giles-Corti, B. and Whitzman, C. (2012) "Active living research: partnerships that count," *Health and Place*, 18(1): 118–20.

Girardi, E., Aloisi, M., Arici, C., Pezzotti, P., Serraino, D., Balzano, R., *et al.* (2004) "Delayed presentation and late testing for HIV: demographic and behavioral risk factors in a multicenter study in Italy," *Journal of Acquired Immune Deficiency Syndrome*, 36(4): 951–9.

GIS Cloud (2014) "GIS Cloud Mobile Data Collection App" [online]. Available at: <http://www.giscloud.com/> (accessed 2 December 2014).

Giskes, K., van Lenthe, F., Avendano-Pabon, M., and Brug, J. (2011) "A systematic review of environmental factors and obesogenic dietary intakes among adults: are we getting closer to understanding obesogenic environments?" *Obesity Reviews*, 12: e95–e106.

Gittelsohn, J., Lee-Kwan, S. H., and Batorsky, B. (2013) "Community-based interventions in prepared-food sources: a systematic review," *Preventing Chronic Disease*, 10. DOI: http://dx.doi.org/10.5888/pcd10.130073.

Gittelsohn, J., Rowan, M., and Gadhoke, P. (2012) "Interventions in small food stores to change the food environment, improve diet, and reduce risk of chronic disease," *Preventing Chronic Disease*, 9. DOI: http://dx.doi.org/10.5888/pcd9.110015.

Gittelsohn, J., Laska, M. N., Karpyn, A., Klingler, K., and Ayala, G. X. (2014) "Lessons learned from small store programs to increase healthy food access," *American Journal of Health Behavior*, 38(2): 307–15.

Glanz, K., Sallis, J. F., Saelens, B. E., and Frank, L. D. (2005) "Healthy nutrition environments: concepts and measures," *American Journal of Health Promotion*, 19(5): 330–3.

Glasgow, R. E., Vogt, T. M., and Boles, S. M. (1999) "Evaluating the public health impact of health promotion interventions: the RE-AIM framework," *American Journal of Public Health*, 89(9): 1322–7.

Gliner, J. A., Morgan, G. A. and Leech, N. L. (2009) *Research Methods in Applied Settings: An Integrated Approach to Design and Analysis*, 2nd edn. New York: Routledge.

Good Food Junction (2014) "Good food junction" [online]. Available at: <http://goodfoodjunction.com/> (accessed 12 December 2014).

Goodman, S., Hammond, D., Pillo-Blocka, F., Glanville, T., and Jenkins, R. (2011) "Use of nutritional information in Canada: national trends between 2004 and 2008," *Journal of Nutrition Education and Behavior*, 43: 356–65.

Gorman, D. M., Speer, P. W., Gruenewald, P. J., and Labouvie, E. W. (2001) "Spatial dynamics of alcohol availability, neighborhood structure and violent crime. *Journal of Studies on Alcohol and Drugs*, 62(5): 628–36.

Government of Ontario (2010) *School Food and Beverage Policy*, Policy/Program Memorandum No. 150 [pdf]. Available at: <http://www.edu.gov.on.ca/extra/eng/ppm/ppm150.pdf> (accessed 15 November 2014).

Green, L. W. (2006) "Public health asks of systems science: to advance our evidence-based practice, can you help us get more practice-based evidence?" *American Journal of Public Health*, 96: 406–9.

Green, L., Poland, B., and Rootman, I. (2000) "The settings approach to health promotion," in B. Poland, L. Green, and I. Rootman (eds), *Settings for Health Promotion: Linking Theory and Practice*. London: Sage, ch. 1.

Green, L. W., Richard, L., and Potvin, L. (1996) "Ecological foundations of health promotion," *American Journal of Health Promotion*, 10(4): 270–81.

Groenewegen, P. P., Van Den Berg, A. E., De Vries, S., and Verheij, R. A. (2006) "Vitamin G: effects of green space on health, well-being, and social safety," *BMC Public Health*, 6(1): 149.

Grunert, K. G., Willis, J. M., and Fernandez-Celemin, L. (2010) "Nutrition knowledge, and use and understanding of nutrition information on food labels among consumers in the UK," *Appetite*, 55(2): 177–89.

Gubler, D. J. (2014) "Dengue viruses: their evolution, history and emergence as a global public health problem," in D. J. Gubler, E. E. Ooi, S. Vasudevan, and J. Farrar (eds), *Dengue and Dengue Hemorrhagic Fever*, 2nd edn. Oxon: CAB International, ch.1.

Gurda, J. and Looze, C. (1999) "The Making of Milwaukee, Milwaukee County Historical Society Milwaukee, WI" [online]. Available at: <http://www.themakingofmilwaukee.com/history/> (accessed 27 November 2015).

Guwani, J. M. and Weech-Maldonado, R. (2004) "Medicaid managed care and racial disparities in AIDS treatment," *Health Care Financing Review*, 26(2): 119–32.

Haan, M. N. and Kaplan, G. A. (1985) *The Contribution of Socioeconomic Position to Minority Health*, Report of the Secretary's Task Force on Black and Minority Health (Vol. 2). Washington, DC: Department of Health and Human Services.

Habluetzel, A., Diallo, D. A., Esposito, F., Lamizana, L., Pagnoni, F., Lengeler, C., *et al.* (1997) "Do insecticide treated curtains reduce all cause child mortality in Burkina Faso?" *Tropical Medicine and International Health*, 2(9): 855–62.

Hadley, J. (2003) "Sicker and poorer – the consequences of being uninsured: a review of the research on the relationship between health insurance, medical care use, health, work, and income," *Medical Care Research and Review*, 60: 3S–75S.

Hadley, J., Holahan, J., Coughlin, T., and Miller, D. (2008) "Covering the uninsured in 2008: current costs, sources of payment, and incremental costs," *Health Affairs*, 27(5): 399–415.

Hagenlocher, M., Delmelle, E., Casas, I., and Kienberger, S. (2013) "Assessing socioeconomic vulnerability to dengue fever in Cali, Colombia: statistical vs expert-based modeling," *International Journal of Health Geographics*, 12(1): 36.

Hall, H. I., McDavid, K., Ling, Q., and Sloggett, A. (2006) "Determinants of progression to AIDS or death after HIV diagnosis, United States, 1996 to 2001," *Annals of Epidemiology*, 16(11): 824–33.

Hall, W. (2009) "The adverse health effects of cannabis use: what are they, and what are their implications for policy?" *International Journal of Drug Policy*, 20: 458–66.

Hall, W. and Degenhardt, L. (2009) "Adverse health effects of non-medical cannabis use," *Lancet*, 17: 1383–91.

Hallal, P. C., Martins, R. C., and Ramírez, A. (2014) "The Lancet Physical Activity Observatory: promoting physical activity worldwide," *The Lancet*, 384(9942): 471–2.

Han, C. S. (2007) "They don't want to cruise your type: gay men of color and the racial politics of exclusion," *Social Identities*, 13(1): 51–67.

Han, C. S. (2008) "A qualitative exploration of the relationship between racism and unsafe sex among Asian Pacific Islander gay men," *Archives of Sexual Behavior*, 37(5): 827–37.

Hanusaik, N., O'Loughlin, J., Kishchuk, N., Paradis, G., and Cameron, R. (2009) "Organizational capacity for chronic disease prevention: a survey of Canadian public health organizations," *European Journal of Public Health*, 20(2): 195–201.

Harrington, D. W. and Elliott, S. J. (2009) "Weighing the importance of neighbourhood: a multilevel exploration of the determinants of overweight and obesity," *Social Science and Medicine*, 68(4): 593–600.

Harrison, K. M., Song, R., and Zhang, X. (2010) "Life expectancy after HIV diagnosis based on national HIV surveillance data from 25 states, United States," *Journal of Acquired Immune Deficiency Syndrome*, 53: 124–30.

Harvest Your Data (2014) "Harvest Your Data" [online]. Available at: <https://www.harvestyourdata.com/> (accessed 2 December 2014).

Hawe, P. (2015) "Lessons from complex interventions to improve health," *Annual Review of Public Health*, 36: 307–23.

Hawe, P. and Potvin, L. (2009) "What is population health intervention research?" *Canadian Journal of Public Health*, 100(1): I8–I13.

Hawe, P., di Ruggiero, E., and Cohen, E. (2012) "Frequently asked questions about population health intervention research," *Canadian Journal of Public Health*, 103(6): e468–e471.

Hawe, P., Shiell, A., and Riley, T. (2004) "Complex interventions: how 'out of control' should a randomized control trial be?" *British Medical Journal*, 328: 1561–3.

Health and Welfare Canada (1987) *Promoting Heart Health in Canada. Report of the Federal-Provincial Working Group on Cardiovascular Disease Prevention and Control.* Ottawa, Canada: Ministry of Supply and Services.

Health Canada (1992) *Heart Health Equality: Mobilizing Communities for Action.* Ottawa: Ministry of Supply and Services.

Health Canada (2013) "Working with grocers to support healthy eating" [online]. Available at: <http://www.hc-sc.gc.ca/fn-an/nutrition/pol/som-ex-sum-grocers-epiciers-eng.php> (accessed 27 November 2015).

Health Resources and Services Administration, HIV/AIDS Bureau (2006) "Outreach: Engaging People in HIV Care" [pdf]. Available at: <ftp://ftp.hrsa.gov/hab/HIVoutreach.pdf > (accessed 27 November 2015).

Healthy Corner Stores Network (2013) "Healthy corner stores network" [online]. Available at: <http://www.healthycornerstores.org> (accessed 27 November 2015).

Healthy Kids Panel (2013) *No Time to Wait: The Healthy Kids Strategy* [pdf]. Available at: <http://www.health.gov.on.ca/en/common/ministry/publications/reports/healthy_kids/healthy_kids.pdf> (accessed 27 August 2013).

Heath, G. W., Parra, D. C., Sarmiento, O. L., Andersen, L. B., Owen, N., Goenka, S., *et al.* (2012) "Evidence-based intervention in physical activity: lessons from around the world," *The Lancet*, 380(9838): 272–81.

Heckert, M. and Mennis, J. (2012) "The economic impact of greening urban vacant land: a spatial difference-in-differences analysis," *Environment and Planning – Part A*, 44(12): 3010–27.

Heikkilä, E. (2005) "Mobile vulnerabilities: perspectives on the vulnerabilities of immigrants in the Finnish labour market," *Population, Space and Place*, 11(6): 485–97.

Hemphill, E. (2008) *State of the Evidence Review on Urban Health and Health Weights*. Ottawa: Canadian Institute for Health Information.

Hemphill, E., Raine, K., Spence, J. C., and Smoyer-Tomic, K. E. (2008) "Exploring obesogenic food environments in Edmonton, Canada: the association between socio-economic factors and fast-food outlet access," *American Journal of Health Promotion*, 22(6): 426–32.

Hewitt, K. (1983) *Interpretations of Calamity from the Viewpoint of Human Ecology*. Boston: Allen & Unwin.

Hii, J. L. K., Smith, T., Vounatsou, P., Alexander, N., Mai, A., Ibam, E., *et al.* (2001) "Area effects of bednet use in a malaria-endemic area in Papua New Guinea," *Transactions of the Royal Society of Tropical Medicine and Hygiene*, 95(1): 7–13.

Hill, J. O. and Peters, J. C. (1998) "Environmental contributions to the obesity epidemic," *Science*, 280: 1371–4.

Hinkel, J. (2011) "Indicators of vulnerability and adaptive capacity: towards a clarification of the science–policy interface," *Global Environmental Change*, 21(1): 198–208.

Ho, L. S., Gittelsohn, J., Rimal, R., Treuth, M. S., Sharma, S., Rosecrans, A., *et al.* (2008) "An integrated multi-institutional diabetes prevention program improves knowledge and healthy food acquisition in Northwestern Ontario first nations," *Health Education and Behavior: The Official Publication of the Society for Public Health Education*, 35(4): 561–73.

Hobin, E. (2010) *Overcoming Barriers to Adopting and Implementing a Province-wide Physical Education Policy for Increasing Adolescent Physical Activity*. Oral presentation at the Canadian Public Health Association Centennial Conference. Toronto.

Hobin, E., Leatherdale, S. T., Manske, S., Dubin, J., Elliott, S., and Veugelers, P. (2012) "A multilevel examination of factors of the school environment and time spent in moderate to vigorous physical activity among a sample of secondary school students in grades 9 to 12 in Ontario, Canada," *International Journal of Public Health*, 57: 699–709.

Hobin, E., So, J., Rosella, L., Comte, M., Manske, S., and McGavock, J. (2014) "Trajectories of objectively measured physical activity among secondary students in Canada in the context of a province-wide physical education policy: a longitudinal analysis," *Journal of Obesity*. DOI: http://dx.doi.org/10.1155/2014/958645.

Hocking, J. S., Rodger, A. J., Rhodes, D. G., and Crofts, N. (2000) "Late presentation of HIV infection associated with prolonged survival following AIDS diagnosis – characteristics of individuals," *International Journal of STD and AIDS*, 11(8): 503–8.

Holsten, J. E. (2009) "Obesity and the community food environment: a systematic review," *Public Health Nutrition*, 12(3): 397–405.

Horton, E. S. (2009) "Effects of lifestyle changes to reduce risks of diabetes and associated cardiovascular risks: results from large- scale efficacy trials," *Obesity*, 17: S43–S48.

Howard, S. C., Omumbo, J., Nevill, C., Some, E. S., Donnelly, C. A., and Snow, R. W. (2000) "Evidence for a mass community effect of insecticide-treated bednets on the incidence of malaria on the Kenyan coast," *Transactions of the Royal Society of Tropical Medicine and Hygiene*, 94(4): 357–60.

Hung, H. C., Joshipura, K. J., and Jiang, R. (2004) "Fruit and vegetable intake and risk of major chronic disease," *Journal of the National Cancer Institute*, 96: 1577–84.

Hunter, D., Perkins, N., Bambra, C., Marks, L., Hopkins, T., and Blackman, T. (2010) *Partnership Working and the Implications for Governance: Issues Affecting Public Health Partnerships*, National Institute for Health Research Service Delivery and Organization Program [pdf]. Available at: <http://www.nets.nihr.ac.uk/__data/assets/pdf_file/0003/64317/FR-08-1716-204.pdf> (accessed 9 April 2016).

Institute for Health Metrics and Evaluation (2010) *GBD Profile: Canada* [pdf]. Available at: <http://www.healthmetricsandevaluation.org/sites/default/files/country-profiles/GBD%20Country%20Report%20-%20Canada.pdf> (accessed 2 April 2013).

Institute of Medicine (2011) *Front-of-Package Nutrition Rating Systems and Symbols: Promoting Healthier Choices* [online]. Available at: <http://www.nap.edu/catalog/13221/front-of-package-nutrition-rating-systems-and-symbols-promoting-healthier> (accessed 2 March 2015).

ISSC and UNESCO (2013) *Changing Global Environments*, World Social Science Report 2013. Paris: OECD and UNESCO Publishing.

Jackson, L. (2014) Environmental Protection Agency, personal communication.

Jacobs, J. (1961) *The Death and Life of Great American Cities*. New York: Random House Digital.

Janssen, I. (2013) "The public health burden of obesity in Canada," *Canadian Journal of Diabetes*, 37: 90–6.

Janssen, I., Boyce, W. F., Simpson, K. and Pickett, W. (2006) "Influence of individual- and area-level measures of socioeconomic status on obesity, unhealthy eating, and physical inactivity in Canadian adolescents," *American Journal of Clinical Nutrition*, 83(1): 139–45.

Jivraj, S. and de Jong, A. (2011) "The Dutch homo-emancipation policy and its silencing effects on queer Muslims," *Feminist Legal Studies*, 19(2): 143–58.

Johnston, R. J., Gregory, D., Pratt, G., and Watts, M. (2000) *The Dictionary of Human Geography*, 4th edn. Malden, MA: Blackwell.

Jones, J., Terashima, M., and Rainham, D. (2009) "Fast food and deprivation in Nova Scotia," *Canadian Journal of Public Health*, 100(1): 32–5.

Jones, N. R., Jones, A., van Sluijs, E., Panter, J., Harrison, F., and Griffin, S. J. (2010) "School environments and physical activity: the development and testing of an audit tool," *Health and Place*, 16: 776–83.

Kaai, S., Brown, K. S., Leatherdale, S. T., Manske, K. S., and Murnaghan, D. (2014) "We do not smoke but some of us are more susceptible than others: a multilevel analysis of a sample of Canadian youth in grades 9 to 12," *Addictive Behaviors*, 39: 1329–36.

Kaai, S., Leatherdale, S. T., Manske, K. S. and Brown, K. S. (2013) "Using student and school factors to differentiate adolescent current smokers from experimental smokers in Canada: a multilevel analysis," *Preventive Medicine*, 57: 113–19.

Kalipeni, E., Craddock, S., Oppong, J. R., and Ghosh J. (2004) *HIV and AIDS in Africa: Beyond Epidemiology*. Malden, MA: Blackwell.

Kania, J. and Kramer, M. (2011) *Collective Impact: Stanford Social Innovation Review* [online]. Available at: <http://www.ssireview.org/articles/entry/collective_impact> (accessed 1 August 2014).

Kaplan, E. L. and Meier, P. (1958) "Nonparametric estimation from incomplete observations," *Journal of the American Statistical Association*, 53(282): 457–81.

Kaplan, S. (1995) "The restorative benefits of nature: toward an integrative framework," *Journal of Environmental Psychology*, 15(3): 169–82.

Katz, M. H., Hsu, L., Lingo, M., Woelffer, G., and Schwarcz, S. K. (1998) "Impact of socioeconomic status on survival with AIDS," *American Journal of Epidemiology*, 148(3): 282–91.

Katzmarzyk, P. T. and Janssen, I. (2004) "The economic costs associated with physical inactivity and obesity in Canada: an update," *Canadian Journal of Applied Physiology*, 29(1): 90–115.

Katzmarzyk, P. T., Reeder, B. A., Elliott, S., Joffres, M. R., Pahwa, P., Raine, K. D., *et al.* (2012) "Body mass index and risk of cardiovascular disease, cancer and all-cause mortality," *Canadian Journal of Public Health*, 103: 147–51.

Kaufmann, C. and Briegel, H. (2004) "Flight performance of the malaria vectors Anopheles gambiae and Anopheles atroparvus" *Journal of Vector Ecology*, 29: 140–53.

Kearns, R. (1993) "Place and health – towards a reformed medical geography," *Professional Geographer*, 45(2): 139–47.

Kearns, R. A. and Joseph, A. (1993) "Space in its place: developing a link in medical geography," *Social Science and Medicine*, 37(6): 711–17.

Kearns, R. and Moon G. (2002) "From medical to health geography: novelty, place and theory after a decade of change," *Progress in Human Geography*, 26(5): 605–25.

Kelly, J. A., Murphy, D. A., Sikkema, K. J., McAuliffe, T. L., Roffman, R. A., Solomon, L. J., *et al.* (1997) "Randomised, controlled, community-level HIV-prevention intervention for sexual-risk behaviour among homosexual men in US cities. Community HIV Prevention Research Collaborative," *The Lancet*, 350(9090): 1500–5.

Kelly, J. A., St Lawrence, J. S., Diaz, Y. E., Stevenson, L. Y., Hauth, A. C., Brasfield, T. L., *et al.* (1991) "HIV risk behavior reduction following intervention with key opinion leaders of population: an experimental analysis," *American Journal of Public Health*, 81(2): 168–71.

Kestens, Y. and Daniel, M. (2010) "Social inequalities in food exposure around schools in an urban area," *American Journal of Preventive Medicine*, 39(1): 33–40.

Kimm, S. Y., Glynn, N. W., Obarzanek, E., Kriska, A. M., Daniels, S. R., Barton, B. A. *et al.* (2005) "Relation between the changes in physical activity and body-mass index during adolescence: a multicentre longitudinal study," *The Lancet*, 366(9482): 301–7.

Kindig, D. and Stoddart, G. (2003) "What is population health?" *American Journal of Public Health*, 93(3): 380–3.

Kirkpatrick, S. I., Reedy, J., Butler, E. N., Dodd, K. W., Subar, A. F., Thompson, F. E., *et al.* (2014) "Dietary assessment in food environment research: a systematic review," *American Journal of Preventive Medicine*, 46(1): 94–102.

Kitahata, M. M., Gange, S. J., Abraham, A. G., Merriman, B., Saag, M. S., Justice, A. C., *et al.* (2009) "Effect of early versus deferred antiretroviral therapy for HIV on survival," *New England Journal of Medicine*, 360(18): 1815–26.

Knigge, L. and Cope, M. (2009) "Grounded visualization and scale: a recursive analysis of community spaces," in M. Cope (ed.), *Qualitative GIS*. London: Sage, Ch. 6.

Koenigstorfer, J. and Klein, A. G. (2013) "Examining the use of nutrition labelling with photoelicitation," *Qualitative Market Research: An International Journal*, 13(4): 389–413.

Kornhauser, R. R. (1978) *Social Sources of Delinquency: An Appraisal of Analytic Models*. Chicago: University of Chicago Press.

Kovats, R. and Butler, C. (2012) "Global health and environmental change: linking research and policy," *Current Opinion in Environmental Sustainability*, 4(1): 44–50.

Kremer, P., Hamstead, Z. A., and McPhearson, T. (2013) "A social-ecological assessment of vacant lots in New York City," *Landscape and Urban Planning*, 120: 218–33.

Krieger, N. (2001a) "A glossary for social epidemiology," *Journal of Epidemiology and Community Health*, 55(10): 693–700.

Krieger, N. (2001b) "Theories for social epidemiology in the 21st century: an ecosocial perspective," *International Journal of Epidemiology*, 30(4): 668–77.

Krieger, N. (2011) *Epidemiology and the People's Health: Theory and Context*. New York: Oxford University Press.

Krieger, N. (2012a) "Methods for the scientific study of discrimination and health: an ecosocial approach," *American Journal of Public Health*, 102(5): 936–44.

Krieger, N. (2012b) "Who and what is a population? Historical debates, current controversies, and implications for understanding population health and rectifying health inequities," *Milbank Quarterly*, 90(4): 634–81.

Krieger, N. (2014) "The real ecological fallacy: epidemiology and global climate change," *Journal of Epidemiology and Community Health*. DOI: 10.1136/jech-2014-205027.

Kuo, F. E. and Sullivan, W. C. (2001) "Environment and crime in the inner city: does vegetation reduce crime?" *Environment and Behavior*, 33(3): 343–67.

Lancet, The (2006) "The US President's malaria initiative," *The Lancet*, 368(9529): 1.

Latham, J. and Moffat, T. (2007) "Determinants of variation in food cost and availability in two socioeconomically contrasting neighbourhoods of Hamilton, Ontario, Canada," *Health and Place*, 13(1): 273–87.

Latkin, C., Mandell, W., Oziemkowska, M., Celentano, D., Vlahov, D., Ensminger, M., *et al.* (1995) "Using social network analysis to study patterns of drug use among urban drug users at high risk for HIV/AIDS," *Drug and Alcohol Dependence*, 38(1): 1–9.

LaVeist, T. A. (2005) Disentangling race and socioeconomic status: a key to understanding health inequalities," *Journal of Urban Health*, 82(2): 26–34.

Le Ber, M. J. and Branzei, O. (2010) (Re)forming strategic cross-sector partnerships: relational processes of social innovation," *Business and Society*, 49(1): 140–72.

Leatherdale, S. T. (2012) "Evaluating school-based tobacco control programs and policies: an opportunity gained and many opportunities lost," *Canadian Journal of Program Evaluation*, 24: 91–108.

Leatherdale, S. T. and Burkhalter, R. (2012) "The substance use profile of Canadian youth: exploring the prevalence of alcohol, drug and tobacco use by gender and grade," *Addictive Behavior*, 37: 318–22.

Leatherdale, S. T. and Papadakis, S. (2011) "A multi-level examination of the association between older social models in the school environment and overweight and obesity among younger students," *Journal of Youth and Adolescence*, 40: 361–72.

Leatherdale, S. T. and Rynard, V. (2013) "A cross-sectional examination of modifiable risk factors for chronic disease among a nationally representative sample of youth: are Canadian students graduating high school with a failing grade for health?" *BMC Public Health*, 13: 569.

Leatherdale, S. T., Bredin, C., and Blashill, J. (2014) "A software application for use in handheld devices to collect school built environment data," *Measurement*, 50: 331–8.

Leatherdale, S. T., Faulkner, G., and Arbour-Nicitopoulos, K. (2010) "School and student characteristics associated with screen-time sedentary behavior among students in grades 5–8, Ontario, Canada, 2007–2008," *Preventing Chronic Diseases: Public Health Research, Practice and Policy (CDC)*, 7: A128.

Leatherdale, S. T., Manske, S., Wong, S., and Cameron, R. (2009) Integrating research, policy and practice in school-based physical activity prevention programming: the School Health Action, Planning and Evaluation System (SHAPES) Physical Activity Module," *Health Promotion Practice*, 10: 254–61.

Leatherdale, S. T., Pouliou, T., Church, D., and Hobin, E. (2011) "The association between overweight and opportunity structures in the built environment: a multi-level analysis among elementary school youth in the PLAY-On study," *International Journal of Public Health*, 56: 237–46.

Leatherdale, S. T., Brown, K. S., Carson, V., Childs, R. A., Dubin, J. A., Elliott, S. J., *et al.* (2014) "The COMPASS study: a longitudinal hierarchical research platform for evaluating natural experiments related to changes in school-level programs, policies and built environment resources," *BMC Public Health*, 14: 331.

Lee, L. M., Karon, J. M., Selik, R., Neal, J. J., and Fleming, P. L. (2001) "Survival after AIDS diagnosis in adolescents and adults during the treatment era, United States, 1984–1997," *Journal of the American Medical Association*, 285(10): 1308.

Leischow, S. J., Best, A., Trochim, W. M., Clark, P. I., Gallagher, R. S., Marcus, S. E., *et al.* (2008) "Systems thinking to improve the public's health," *American Journal of Preventive Medicine*, 35: s196–s203.

Lerner, R. M., Ostrom, C. W., and Freel, M. A. (1997) "Preventing health-compromising behaviors among youth and promoting their positive development: a developmental contextual perspective," in J. Schulenberg and K. Hurrelmann (eds), *Health Risks and Developmental Transitions During Adolescence*. New York: Cambridge University Press, ch. 19.

Lévesque, L., Guilbault, G., Delormier, T., and Potvin, L. (2005) "Unpacking the black box: a deconstruction of the programming approach and physical activity interventions implemented in the Kahnawake Schools Diabetes Prevention Project," *Health Promotion Practice*, 6(1): 64–71.

Lewis, N. M. (2014) "Rupture, resilience, and risk: relationships between mental health and migration among gay-identified men in North America," *Health and Place*, 27: 212–19.

Liggett, R., Loukaitou-Sideris, A., and Iseki, H. (2001) "Bus stop-environment connection: do characteristics of the built environment correlate with bus stop crime?" *Transportation Research Record: Journal of the Transportation Research Board*, 1760(1): 20–7.

Lincoln, Y. and Guba, E. (2003) *Naturalistic Inquiry*. Beverly Hills, CA: Sage.

Lindsay, J. R. (2003) "The determinants of disaster vulnerability: achieving sustainable mitigation through population health," *Natural Hazards*, 28(2–3): 291–304.

Litva, A. and Eyles, J. (1995) "Coming out: exposing social theory in medical geography," *Health and Place*, 1(1): 5–14.

Lozano, R., Naghavi, M., Foreman, K., Lim, S., Shibuya, K., *et al.* (2012) "Global and regional mortality from 235 causes of death for 20 age groups in 1990 and 2010: a systematic analysis for the Global Burden of Disease Study 2010," *The Lancet*, 380(9859): 2095–128.

Luers, A. (2005) "The surface of vulnerability: an analytical framework for examining environmental change," *Global Environmental Change*, 15(3): 214–23.

Luginaah, I. (2009) "Health geography in Canada: where are we headed?" *The Canadian Geographer/Le Géographe Canadien*, 53(1): 91–9.

Luo, W., Morrison, H., de Groh, M., Waters, C., DesMeules, M., Jones-McLean, E., *et al.* (2007) "The burden of adult obesity in Canada," *Chronic Diseases in Canada*, 27(4): 135–44.

Lytle, L. A. (2009) "Measuring the food environment: state of the science," *American Journal of Preventive Medicine*, 36(4 Suppl): S134–44.

Maas, J., Verheij, R. A., Groenewegen, P. P., de Vries, S., and Spreeuwenberg, P. (2006) "Green space, urbanity, and health: how strong is the relation?" *Journal of Epidemiology and Community Health*, 60(7): 587–92.

McCarthy, J. J., Canziani, O. F., Leary, N. A., Dokken, D. J., and White, K. S. (eds), (2001) *Climate Change: Impacts, Adaptation and Vulnerability*, Intergovernmental Panel on Climate Change, Working Group II. Cambridge: Cambridge University Press.

McCormack, G. R., Giles-Corti, B., and Bulsara, M. (2008) "The relationship between destination proximity, destination mix and physical activity behaviors," *Preventive Medicine*, 46(1): 33–40.

Macintyre, S. and Ellaway, A. (2000) "Ecological approaches: rediscovering the role of the physical and social environment," in L. F. Berkman and I. Kawachi (eds), *Social Epidemiology*. New York: Oxford University Press, ch. 14.

Macintyre, S., Ellaway, A., and Cummins, S. (2002) "Place effects on health: how can we conceptualise, operationalise, and measure them?" *Social Science and Medicine*, 55(1): 125–39.

McKenna, M. L. (2010) "Policy options to support healthy eating in schools," *Canadian Journal of Public Health*, 101(Suppl. 2): S14–S17.

McLaren, L. and Emery, J. C. H. (2012) "Drinking water fluoridation and oral health inequities in Canadian children," *Canadian Journal of Public Health*, 103(7): eS49–eS56.

McLaughlin, P. and Dietz, T. (2008) "Structure, agency and environment: toward an integrated perspective on vulnerability," *Global Environmental Change*, 18(1): 99–111.

McLeroy, K. R., Bibeau, D., Steckler, A., and Glanz, K. (1988) "An ecological perspective on health promotion programs," *Health Education Quarterly*, 15(4): 351–77.

McMichael, A. (1999) "Prisoners of the proximate: loosening the constraints on epidemiology in an age of change," *American Journal of Epidemiology*, 149(10): 887–97.

McMichael, A. (2006) "Population health as the 'bottom line' of sustainability: a contemporary challenge for public health researchers," *European Journal of Public Health*, 16(6): 579–81.

McMichael, A. (2014) "Global change and human health: introduction," *Global Environmental Change, Handbook of Global Environmental Pollution*, 1: 599–603.

McMichael, A. J., Campbell-Lendrum, D. H., Corvalán, C. F., Ebi, K. L., Githeko, A. K., Scheraga, J. D., *et al.* (2003) *Climate Change and Human Health*. Geneva: World Health Organization.

Magesa, S. M., Wilkes, T. J., Mnzava, A. E. P, Njunwa, K. J., Myamba, J., Kivuyo, M. D. P., *et al.* (1991) "Trial of pyrethroid impregnated bednets in an area of Tanzania holoendemic for malaria Part 2. Effects on the malaria vector population," *Acta Tropica*, 49(2): 97–108.

Manitoba Education, Citizenship, and Youth (2007) *Implementation of Grades 11 and 12 Physical Education/Health Education: A Policy Document* [pdf]. Available at: <http://www.edu.gov.mb.ca/k12/docs/policy/imp_pehe/document.pdf > (accessed 27 November 2015).

Margulis, H. L. and Kenny, J. T. (2001) "When city and country collide – managing growth in the metropolitan fringe. Tom Daniels: The Making of Milwaukee. John Gurda," *Urban Geography*, 22(3): 287–90.

Marks, G., Crepaz, N., and Janssen, R. S. (2006) "Estimating sexual transmission of HIV from persons aware and unaware that they are infected with the virus in the USA," *AIDS*, 20(10): 1447–50.

Marmot, M. (2007) "Achieving health equity: from root causes to fair outcomes," *The Lancet*, 370(9593): 1153–63.

Marmot, M., Friel, S., Bell, R., Houweling, T. A., and Taylor, S. (2008) "Closing the gap in a generation: health equity through action on the social determinants of health," *The Lancet*, 372(9650): 1661–9.

Martinez, O., Wu, E., Sandfort, T., Shultz, A. Z., Capote, J., Chavez, S., *et al.* (2014) "A couple-based approach: an innovative effort to tackle HIV infection among Latino gay men," *EHQUIDAD*, 1: 15–32.

Mâsse, L. C. and de Niet, J. E. (2013) "School nutrition capacity, resources and practices are associated with availability of food/beverage items in schools," *International Journal of Behavioral Nutrition and Physical Activity*, 10: 26.

Mâsse, L. C., Naiman, D., and Naylor, P.-J. (2013) From policy to practice: implementation of physical activity and food policies in schools," *International Journal of Behavioral Nutrition and Physical Activity*, 10: 71–82.

Masuda, J. R., Robinson, K., Elliott, S. J., and Eyles J. (2012) "Chronic disease prevention and the politics of scale: lessons from Canadian Health Reform," *Social Work in Public Health*, 27(7): 639–57.

Masuda, J. R., Robinson, K., Elliott, S. J., and Eyles, J. (2009) "Disseminating chronic disease prevention 'to or with' Canadian public health systems," *Health Education and Behavior*, 36(6): 1026–50.

Meade, M. S. (1977) "Medical geography as human ecology: the dimension of population movement," *Geographical Review*, 67(4): 379–93.

Meditz, A. L., MaWhinney, S., Allshouse, A., Feser, W., Markowitz, M., Little, S., *et al.* (2011) "Sex, race, and geographic region influence clinical outcomes following primary HIV-1 infection," *Journal of Infectious Diseases*, 203(4): 442–51.

Melendez, R. M., Zepeda, J., Samaniego, R., Chakravarty, D., and Alaniz, G. (2013) "'La Familia' HIV prevention program: a focus on disclosure and family acceptance for Latino immigrant MSM to the USA," *Salud Pública de México*, 55(Suppl. 4): S491–S497.

Mercille, G., Richard, L., Gauvin, L., Kestens, Y., Payette, H., and Daniel, M. (2013) "Comparison of two indices of availability of fruits/vegetable and fast food outlets," *Journal of Urban Health*, 90(2): 240–5.

Messina, J. P., Taylor, S. T., Meshnick, S. R., Linke, A. M., Tshefu, A. K., Atua, B., *et al.* (2011) "Population, behavioural and environmental drivers of malaria prevalence in the Democratic Republic of Congo," *Malaria Journal*, 10: 161.

Miles, M. B. and Huberman, A. M. (1994) *Qualitative Data Analysis: An Expanded Sourcebook*, 2nd edn. Thousand Oaks, CA: Sage.

Miller, F., Osbahr, H., Boyd, E., Thomalla, F., Bharwani, S., Ziervogel, G., *et al.* (2010) "Resilience and vulnerability: complementary or conflicting concepts," *Ecology and Society*, 15(3): 11.

Miller, J. W., Naimi, T. S., Brewer, R. D. and Jones, S. E. (2007) "Binge drinking and associated health risk behaviors among high school students," *Pediatrics*, 119: 76–85.

Minaker, L. (2013) *Measuring the Food Environment in Canada*. Ottawa: Health Canada.

Minkler, M. (2000) "Using participatory action research to build healthy communities," *Public Health Reports*, 115(2–3): 191–7.

Mohammad, A., Emch, M., von Seidlein, L., Yunus, M., Sack, D. A., Rao, M., *et al.* (2005) "Herd immunity conferred by killed oral cholera vaccines in Bangladesh: a reanalysis," *The Lancet*, 366(9479): 44–9.

Mohammad, A., Emch, M., Yunus, M., Sack, D., Lopez, A.L., Holmgren, J., *et al.* (2008) "Vaccine protection of Bangladeshi infants and young children against cholera: implications for vaccine deployment and person-to-person transmission," *Pediatric Infectious Disease Journal*, 27(1): 33–7.

Monteiro, C. A. (2009) "All the harmful effects of ultra-processed foods are not captured by nutrient profiling," *Public Health Nutrition*, 12(10): 1968–9.

Monteiro, C. A., Moubarac, J., Cannon, G., Ng, S. W., and Popkin, B. (2013) "Ultra-processed products are becoming dominant in the global food system," *Obesity Reviews*, 14(S2): 21–8.

Montgomery, J. P., Gillespie, B. W., Gentry, A. C., Mokotoff, E. D., Crane, L. R., and James, S. A. (2002) "Does access to health care impact survival time after diagnosis of AIDS?" *AIDS Patient Care and STDs*, 16(5): 223–31.

Moon, G., Quarendon, G., Barnard, S., Twigg, L., and Blyth, B. (2007) "Fat nation: deciphering the distinct geographies of obesity in England," *Social Science and Medicine*, 65(1): 20–31.

Moore, L. V. and Diez Roux, A. V. (2006) "Associations of neighborhood characteristics with the location and type of food stores," *American Journal of Public Health*, 96(2): 325–31.

Morenoff, J. D., Sampson, R. J., and Raudenbush, S. W. (2001) "Neighborhood inequality, collective efficacy, and the spatial dynamics of urban violence," *Criminology*, 39(3): 517–58.

Morens, D. M. and Fauci, A. S. (2014) "Chikungunya at the door – déjà vu all over again?" *New England Journal of Medicine*, 371(10): 885–7.

Moubarac, J., Martins, A. P. B., Claro, R. M., Levy, R. B., Cannon, G., and Monteiro, C. A. (2012) "Consumption of ultra-processed foods and likely impact on human health. Evidence from Canada," *Public Health Nutrition*, 1(1): 1–9.

Mullaly, M. L., Taylor, J. P., Kuhle, S., Bryanton, J., Hernandez, K. J., MacLellan, D. L., *et al.* (2010) "A province-wide school nutrition policy and food consumption in elementary school children in Prince Edward Island," *Canadian Journal of Public Health*, 101(1): 40–3.

Mwangi, T. W., Ross, A., Marsh, K., and Snow, R. W. (2003) "The effects of untreated bednets on malaria infection and morbidity on the Kenyan coast," *Transactions of the Royal Society of Tropical Medicine and Hygiene*, 97(4): 369–72.

Myers, S. and Patz, J. (2009) "Emerging threats to human health from global environmental change," *Annual Review of Environment and Resources*, 34(1): 223–52.

Nahlen, B. L., Clark, J. P., and Alnwick, D. (2003) "Insecticide-treated bed nets," *American Journal of Tropical Medicine and Hygiene*, 68(4 suppl.): 1–2.

Nash, C. J. (2013) "The age of the 'Post-mo'? Toronto's gay village and a new generation," *Geoforum*, 49: 243–52.

National Cancer Institute (2014) "Measures of the food environment" [online]. Available at: <http://appliedresearch.cancer.gov/mfe/> (accessed 27 November 2015).

Neumark-Sztainer, D., French, S. A., Hannan, P. J., Story, M., and Fulkerson, J. A. (2005) "School lunch and snacking patterns among high school students: association with school food environment and policies," *International Journal of Behavioral Nutrition and Physical Activity*, 2: 14–20.

Nevill, C. G., Some, E. S., Mung'Ala, V. O., Muterni, W., New, L., Marsh, K., *et al.* (1996) "Insecticide-treated bednets reduce mortality and severe morbidity from malaria among children on the Kenyan coast," *Tropical Medicine and International Health*, 1(2): 139–46.

New York City (2013) "Food retail expansion to support health" [online]. Available at: <http://www.nyc.gov/html/misc/html/2009/fresh.shtml> (accessed 12 December 2014).

Ni Mhurchu, C., Vandevijvere, S., Waterlander, W., Thornton, L. E., Kelly, B., Cameron, A. J., *et al.* (2013) "Monitoring the availability of healthy and unhealthy foods and non-alcoholic beverages in community and consumer retail food environments globally," *Obesity Reviews*, 14(S1): 108–19.

Nutbeam, D. (1996) "Health outcomes and health promotion: defining success in health promotion," *Health Promotion Journal of Australia*, 6: 58–60.

O'Brien, K., Eriksen, S., Nygaard, L., and Schjolden, A. (2007) "Why different interpretations of vulnerability matter in climate change discourses," *Climate Policy*, 7(1): 73–88.

Ogilvie, D., Griffin, S., Jones, A., Mackett, R., Guell, C., Panter, J., *et al.* (2010) "Commuting and health in Cambridge: a study of a "natural experiment" in the provision of new transport infrastructure," *BMC Public Health*, 10(1): 703.

Ontario Ministry of Health and Long-Term Care (2013) *No Time to Wait: The Healthy Kids Strategy. Healthy Kids Panel* [pdf]. Ottawa: Queen's Printer for Ontario. Available at: <http://www.health.gov.on.ca/en/common/ministry/publications/reports/healthy_kids/healthy_kids.pdf> (accessed 10 February 2015).

Oppong, J. R. and Harold, A. (2009) "Disease, ecology, and environment," in T. Brown, S. McLafferty, and G. Moon (eds), *A Companion to Health and Medical Geography*. Chichester: Wiley-Blackwell, ch. 5.

Oppong, J. R., Kutch, L., Tiwari, C., and Arbona, S. (2014) "Vulnerable places: prison locations, socioeconomic status and HIV infection in Texas," *Professional Geographer*, 66(4): 653–63.

Oppong, J. R., Tiwari, C., Ruckthongsook, W., Huddleston, J., and Arbona, S. (2012) "Mapping late testers for HIV in Texas," *Health and Place*, 18: 568–75.

Pabayo, R., O'Loughlin, J., Gauvin, L., Paradis, G., and Gray-Donald, K. (2006) "Effect of a ban on extracurricular sports activities by secondary school teachers on physical activity levels of adolescents: a multilevel analysis," *Health Education and Behavior*, 33(5): 690–702.

Palella, F. J., Deloria-Knoll, M., Chmiel, J. S., Moorman, A. C., Wood, K. C., Greenberg, A. E., *et al.* (2003) "Survival benefit of initiating antiretroviral therapy in HIV-infected persons in different CD4+ cell strata," *Annals of Internal Medicine*, 138(8): 620–6.

Papas, M. A., Alberg, A. J., Ewing, R., Helzlsouer, K. J., Gary, T. L., and Klassen, A. C. (2007) "The built environment and obesity," *Epidemiological Reviews*, 29: 129–43.

Parcel, G. S., Kelder, S. H., and Basen-Engquist, K. (2000) "The school as a setting for youth health promotion," in B. Poland, L. W. Green, and I. Rootman (eds), *Settings for Health Promotion: Linking Theory and Practice*. Thousand Oaks, CA: Sage.

Park, I. K. and Ciorici, P. (2013) "Determinants of vacant lot conversion into community gardens: evidence from Philadelphia," *International Journal of Urban Sciences*, 17(3): 385–98.

Parkes, M. and Horwitz, P. (2009) "Water, ecology and health: ecosystems as settings for promoting health and sustainability," *Health Promotion International*, 24(1): 94–102.

Pawson, R. (2013) *The Science of Evaluation: A Realist Manifesto*. London: Sage.

Pearce, J., Blakely, T., Witten, K., and Bartie, P. (2007) "Neighborhood deprivation and access to fast-food retailing," *American Journal of Preventive Medicine*, 32(5): 375–82.

Peel, J. L., Klein, M., Flanders, W. D., Mulholland, J. A., and Tolbert, P. E. (2008) "Impact of improved air quality during the 1996 Atlanta Summer Olympic Games on cardiovascular and respiratory outcomes," *Epidemiology*, 19(6): S318.

Pennsylvania Horticultural Society (2014) *LandCare Program* [online]. Available at: <http://phsonline.org/index.php/greening/landcare-program> (accessed 15 May 2014).

Perez-Heydrich, C., Hudgens, M. G., Halloran, M. E., Clemens, J. D., Ali, M., and Emch, M. E. (2014) "Assessing effects of cholera vaccination in the presence of interference," *Biometrics*, 70(3): 731–44.

Petticrew, M., Cummins, S., Ferrell, C., Findlay, A., Higgins, C., Hoy, C., *et al.* (2005) "Natural experiments: an underused tool for public health?" *Public Health*, 119: 751–7.

Petticrew, M., Chalabi, Z., and Jones, D. R. (2012a) "To RCT or not to RCT: deciding when 'more evidence is needed' for public health policy and practice," *Journal of Epidemiology and Community Health*, 66(5): 391–6.

Petticrew, M., Tugwell, P., Kristjansson, E., Oliver, S., Ueffing, E., and Welch, V. (2012b) "Damned if you do, damned if you don't: subgroup analysis and equity," *Journal of Epidemiology and Community Health*, 66(1): 95–8.

Phills, J. A., Deiglmeier, K., and Miller, D. T. (2008) "Rediscovering social innovation," *Stanford Social Innovation Review*, 6: 34–43.

Plummer, R., de Loë, R., and Armitage, D. (2012) "A systematic review of water vulnerability assessment tools," *Water Resources Management*, 26(15): 4327–46.

Poland, B. (1993) *Concept and Practice in Community Mobilization for Health: A Qualitative Evaluation of the Brantford COMMIT Smoking Cessation Intervention Trial*. Doctoral dissertation, McMaster University.

Poland, B., Frohlich, K., and Cargo, M. (2008) "Context as a fundamental dimension of health promotion program evaluation," in L. Potvin and D. McQueen (eds), *Health Promotion Evaluation Practices in the Americas*. New York: Springer, ch. 17.

Poland, B., Coburn, D., Robertson, A., Eakin, J., and members of the Critical Social Science Group (1998) "Wealth, equity and health care: a critique of a 'population health' perspective on the determinants of health," *Social Science and Medicine*, 46(7): 785–98.

Policy Link (2014) *The Healthy Food Financing Initiative (HFFI): An Innovative Public-Private Partnership Sparking Economic Development and Improving Health* [pdf]. Available at: <http://www.healthyfoodaccess.org/sites/default/files/HFFI%20Fact%20Sheet%2010%2016%2014.pdf> (accessed 12 December 2014).

Polsky, J. Y., Moineddin, R., Glazier, R. H., Dunn, J. R., and Booth, G. L. (2014) "Foodscapes of southern Ontario: neighbourhood deprivation and access to healthy and unhealthy food retail," *Canadian Journal of Public Health*, 105(5): e369–e375.

Pope, C., Ziebland, S., and Mays, N. (2000) "Qualitative research in health care: analysing qualitative data," *British Medical Journal*, 320: 114–16.

Potvin, L. (2012) "A critical look at a nascent field," *Canadian Journal of Public Health*, 103(Suppl. 1): S63–S64.

Potvin, L. (2013) "Population health intervention research: a fundamental science for NCD prevention," in D. V. Mcqueen (ed.), *Global Handbook on Noncommunicable Disease and Health Promotion*. New York: Springer, ch. 13.

Potvin, L. and Lamarre, M.-C. (2009) "Entre politiques nationales et initiatives locales, renforcer les synergies pour promouvoir la santé," *Global Health Promotion*, 16(4): 75–82.

Potvin, L., Di Ruggiero, E., and Shoveller, J. A. (2013) "Pour une science des solutions: la recherche interventionnelle en santé des populations," *La Santé en Action*, 423: 13–15.

Potvin, L., Gendron, S., and Bilodeau, A. (2012) "Trois conceptions de la nature des programmes: implications pour l'évaluation de programmes complexes en santé publique," *La Revue Canadienne d'Évaluation de Programme*, 26(3): 91–104.

Potvin, L., Haddad, S., and Frohlich K. L. (2001) "Beyond process evaluation," in I. Rootman, M. Goodstadt, B. Hyndman, D. V. Mcqueen, L. Potvin, J. Springett, *et al.* (eds), *Evaluation in Health Promotion. Principles and Perspectives*. Copengagen: WHO Regional Publications, European Series, No. 92, Ch. 2.

Potvin, L., Gendron, S., Bilodeau, A., and Chabot, P. (2005) "Integrating social science theory into public health practice," *American Journal of Public Health*, 95(2): 91–104.

Pouliou, T. and Elliott, S. J. (2010) "Individual and socio-environmental determinants of overweight and obesity in urban Canada," *Health and Place*, 16: 389–98.

Poznansky, M. C., Coker, R., Skinner, C., Hill, A., Bailey, S., Whitaker, L., *et al.* (1995) "HIV positive patients first presenting with an AIDS defining illness: characteristics and survival," *British Medical Journal*, 311(6998): 156–8.

Preston, B., Yuen, E., and Westaway, R. (2011) "Putting vulnerability to climate change on the map: a review of approaches, benefits, and risks," *Sustainability Science*, 6(2): 177–202.

Provan, K. G. and Lemaire, R. (2012) "Core concepts and key ideas for understanding public sector organizational networks: using research to inform scholarship and practice," *Public Administration Review*, 72(5): 638–48.

Provan, K. G., Fish, A., and Sydow, J. (2007) "Interorganizational networks at the network level: a review of the empirical literature on whole networks," *Journal of Management*, 33(3): 479–516.

Public Health Agency of Canada (2010) *Curbing Childhood Obesity: A Federal, Provincial and Territorial Framework for Action to Promote Healthy Weights*. Ottawa: Public Health Agency of Canada.

Public Health Agency of Canada (2011) *Obesity in Canada: A Joint Report from the Public Health Agency of Canada and the Canadian Institute for Health Information*. Ottawa: Public Health Agency of Canada.

Public Health Agency of Canada (2012) *What Is the Population Health Approach?* [online]. Available at: <http://www.phac-aspc.gc.ca/ph-sp/approach-approche/index-eng.php> (accessed 18 June 2015).

Public Health Agency of Canada (2013) *School Health* [online]. Available at: <http://www.phac-aspc.gc.ca/hp-ps/dca-dea/prog-ini/school-scolaire/index-eng.php> (accessed 1 December 2015).

Public Health Agency of Canada (2014) *Multi-Sectoral Partnerships to Promote Healthy Living and Prevent Chronic Disease* [online]. Available at: <http://www.phac-aspc.gc.ca/fo-fc/mspphl-pppmvs-eng.php> (accessed 23 May 2014).

Public Health Agency of Canada (2015) *Social Determinants of Health* [online]. Available at: <http://cbpp-pcpe.phac-aspc.gc.ca/public-health-topics/social-determinants-of-health/> (accessed 24 June 2015).

Pucher, J., Dill, J., and Handy, S. (2010) "Infrastructure, programs, and policies to increase bicycling: an international review," *Preventive Medicine*, 50: S106–S125.

Québec en Forme (2012) *Modifying the Built Environment to Promote Healthy Eating Among Youth* [pdf]. Available at: <http://www.quebecenforme.org/media/103607/08_research_summary.pdf> (accessed 12 December 2014).

Quintanilha, M., Downs, S., Lieffers, J., Berry, T., Farmer, A., and McCargar, L. J. (2013) "Factors and barriers associated with early adoption of nutrition guidelines in Alberta, Canada," *Journal of Nutrition Education and Behavior*, 45(6): 510–17.

Raine, K. D., Spence, J. D., Church, J., Boule, N., Slater, L., Marko, J., *et al.* (2008) *State of the Evidence Review on Urban Health and Health Weights*. Ottawa: Canadian Institute for Health Information.

Ramanathan, S., Allison, K., Faulkner, G., and Dwyer, J. (2008) "Challenges in assessing the implementation and effectiveness of physical activity and nutrition policy interventions as natural experiments," *Health Promotion International*, 23: 290–7.

Ramirez-Valles, J. (2007) "The quest for effective HIV-prevention interventions for Latino gay men," *American Journal of Preventive Medicine*, 32(4S): S34–S35.

Rapiti, E., Porta, D., Forastiere, F., Fusco, D., and Perucci, C. A. (2000) "Socioeconomic status and survival of persons with AIDS before and after the introduction of highly active antiretroviral therapy," *Epidemiology*, 11(5): 496–501.

RBM (2005) *Global Strategic Plan. Roll Back Malaria 2005–2015*. Geneva: Roll Back Malaria Partnership.

Regis, L., Acioli, R., Silveira, J., Melo-Santos, M., Souza, W., Ribeiro, C., *et al.* (2013) "Sustained reduction of the dengue vector population resulting from an integrated control strategy applied in two Brazilian cities," *PLoS ONE*, 8(7): e67682.

Rehm, J., Gmel, G., Sempos, C. T., and Trevisan, M. (2002) "Alcohol-related morbidity and mortality," *Alcohol Research and Health*, 27: 39–51.

Rhodes, S. D., Daniel, J., Alonzo, J., Duck, S., Garcia, M., Downs, M., *et al.* (2013) "A systematic community-based participatory approach to refining an evidence-based community-level intervention: the HOLA intervention for Latino men who have sex with men," *Health Promotion Practice*, 14(4): 607–16.

Richard, L., Gauvin, L., and Raine, K. (2011) "Ecological models revisited: their uses and evolution in health promotion over two decades," *Annual Review of Public Health*, 32: 307–26.

Riley, B. and Feltracco, A. (2002) *Situational Analysis of the Canadian Heart Health Initiative: Final Report*. Ottawa: Health Canada.

Ringwalt, C., Ennett, S. T., Vincus, A. A., Rohrback, L. A., and Simons-Rudolph, A. (2004) "Who's calling the shots? Decision-makers and the adoption of effective school-based substance use prevention curricula," *Journal of Drug Education*, 34: 19–31.

Roberts, K. C., Shields, M., de Groh, M., Aziz, A., and Gilbert, J. A. (2012) "Overweight and obesity in children and adolescents: results from the 2009 to 2011 Canadian Health Measures Survey," *Health Reports*, 23(3): 37–41.

Robinson, K. and Elliott, S. J. (2000) "The practice of community development approaches in heart health promotion," *Health Education Research*, 15(2): 219–31.

Robinson, K., Farmer, T., Elliott, S. J., Eyles, J. (2007a) "From heart health promotion to chronic disease prevention: contributions of the Canadian Heart Health Initiative," *Preventing Chronic Disease, Public Health Research, Practice, and Policy*, 4(2): A29.

Robinson, K., Farmer, T., Riley, B., Elliot, S. J., and Eyles, J. (2007b) "Realistic expectations: investing in organizational capacity building for chronic disease prevention," *American Journal of Health Promotion*, 21(5): 430–8.

Rockström, J., Steffen, W., Noone, K., Persson, Å., Chapin, F., Lambin, E., *et al.* (2009) "A safe operating space for humanity," *Nature*, 461(7263): 472–5.

Rogers, E. M. (2003) *Diffusion of Innovations*, 5th edn. New York: Free Press.

Rose, G. (1985) "Sick individuals and sick populations," *International Journal of Epidemiology*, 14: 32–8.

Rose, D., Bodor, J. N., Hutchinson, P. L., and Swalm, C. M. (2010) "The importance of a multi-dimensional approach for studying the links between food access and consumption," *Journal of Nutrition*, 140(6): 1170–4.

Rosecrans, A. M., Gittelsohn, J., Ho, L. S., Harris, S. B., Naqshbandi, M., and Sharma, S. (2008) "Process evaluation of a multi-institutional community-based program for diabetes prevention among first nations," *Health Education Research*, 23(2): 272–86.

Ross, C. E. and Mirowsky, J. (1999) "Disorder and decay the concept and measurement of perceived neighborhood disorder," *Urban Affairs Review*, 34(3): 412–32.

Ross, N. A. and Taylor, S. M. (1998) "Geographical variation in attitudes towards smoking: findings from the COMMIT communities," *Social Science and Medicine*, 46(6): 703–17.

Rosser, B. R. S., West, W., and Weinmeyer, R. (2008) "Are gay communities dying or just in transition? Results from an international consultation examining possible structural change in gay communities," *AIDS Care*, 20(5): 588–95.

Rychetnik, L., Frommer, M., Hawe, P., and Shiell, A. (2002) "Criteria for evaluating evidence on public health interventions," *Journal of Epidemiology and Community Health*, 56(2): 119–27.

St John, M., Durant, M., Campagna, P. D., Rehman, L. A., Thompson, A. M., Wadsworth, L. A., *et al.* (2008) "Overweight Nova Scotia children and youth: the roles of household income and adherence to Canada's food guide to healthy eating," *Canadian Journal of Public Health*, 99(4): 301–6.

Sampson, R. J. (2009) "Disparity and diversity in the contemporary city: social (dis)order revisited," *British Journal of Sociology*, 60(1): 1–31.

Sampson, R. J. and Groves, W. B. (1989) "Community structure and crime: testing social-disorganization theory," *American Journal of Sociology*, 94(4): 774–802.

Sampson, R. J., Morenoff, J. D., and Gannon-Rowley, T. (2002) "Assessing neighborhood effects: social processes and new directions in research," *Annual Review of Sociology*, 28: 443–78.

Sampson, R. J., Morenoff, J. D., and Raudenbush, S. (2005) "Social anatomy of racial and ethnic disparities in violence," *American Journal of Public Health*, 95(2): 224–32.

Sanez, R. (2005) *The Social and Economic Isolation of Urban African Americans* [online]. Available at: <http://www.prb.org/Articles/2005/TheSocialandEconomicIsolationof UrbanAfricanAmericans.aspx> (accessed 18 December 2010).

Schackman, B. R., Gebo, K. A., Walensky, R. P., Losina, E., Muccio, T., Sax, P. E., *et al.* (2006) "The lifetime cost of current human immunodeficiency virus care in the United States," *Medical Care*, 44(11): 990–7.

Schön, D. A. and Rein, M. (1994) *Frame Reflection: Toward the Resolution of Intractable Policy Controversies*. New York: Basic Books.

Schulz, A. J., Williams, D. R., Israel, B. A., and Lempert, L. B. (2002) "Racial and spatial relations as fundamental determinants of health in Detroit," *Milbank Quarterly*, 80(4): 677–707.

Schwab, M. S. and Syme, L. (1997) "On paradigms, community participation, and the future of public health," *American Journal of Public Health*, 87(12): 2049–51.

Seitanidi, M. M., Koufopoulos, D. N., and Palmer, P. (2010) "Partnership formation for change: indicators for transformative potential in cross sector social partnerships," *Journal of Business Ethics*, 94(S1): 139–61.

Seliske, L. M., Pickett, W., Boyce, W. F., and Janssen, I. (2009) "Density and type of food retailers surrounding Canadian schools: variations across socioeconomic status," *Health and Place*, 15(3): 903–7.

Sen, A. (1981) "Ingredients of famine analysis: availability and entitlements," *Quarterly Journal of Economics*, 96(3): 433–64.

Shadish, W. R. and Cook, T. D. (2009) "The renaissance of field experimentation in evaluating interventions," *Annual Review of Psychology*, 60(1): 607–29.

Shaheen, S. A., Guzman, S., and Zhang, H. (2010) "Bikesharing in Europe, the Americas, and Asia," *Transportation Research Record: Journal of the Transportation Research Board*, 2143: 159–67.

Shelley, J. J. (2012) "Addressing the policy cacophony does not require more evidence: an argument for reframing obesity as caloric overconsumption," *BMC Public Health*, 12: 1042.

Shields, M. and Tremblay, M. (2008) "Sedentary behaviour and obesity," *Health Reports*, 19: 19–30.

Shoveller, J., Knight, R., Thompson, K., and Greyson, D. (2015) "Going beyond 'context matters' to propose how and why context influences public health interventions," *European Journal of Public Heath*, 25(Suppl. 3). DOI: 10.1093/eurpub/ckv167.028.

Singh, G. K., Azuine, R. E., and Siahpush, M. (2013) "Widening socioeconomic, racial, and geographic disparities in HIV/AIDS mortality in the United States, 1987–2011," *Advances in Preventive Medicine*. DOI: 10.1155/2013/657961.

Smith, K. E., Bambra, C., Joyce, K. E., Perkins, N., Hunter, D. J., and Blenkinsopp, E. A. (2009) "Partners in health? A systematic review of the impact of organizational partner-ships on public health outcomes in England between 1997 and 2008," *Journal of Public Health*, 31(2): 210–21.

Smith, T. and Smith, B. (2004) *Kaplan Meier and Cox Proportional Hazards Modeling: Hands on Survival Analysis* [pdf]. Available at: <http://www.lexjansen.com/wuss/2004/hands_on_workshops/i_how_kaplan_meier_and_cox_p.pdf> (accessed 7 March 2013).

Smoyer-Tomic, K. E., Spence, J. C., Raine, K. D., Amrhein, C., Cameron, N., Yasenovskiy, V., *et al.* (2008) "The association between neighborhood socioeconomic status and exposure to supermarkets and fast food outlets," *Health and Place*, 14(4): 740–54.

Snijders, T. A. B. and Bosker, R. J. (2012) *Multilevel Analysis: An Introduction to Basic and Advanced Multilevel Modeling*, 2nd edn. London: Sage, Ch. 15.

Snow, R. W., Craig, M., Deichmann, U., and Marsh, K. (1999) "Estimating mortality, morbidity and disability due to malaria among Africa's non-pregnant population," *Bulletin of the World Health Organization*, 77(8): 624.

Snow, R. W., Rowan, K. M., Lindsay, S. W., and Greenwood, B. M. (1988) "A trial of bed nets (mosquito nets) as a malaria control strategy in a rural area of the Gambia, West Africa," *Transactions of the Royal Society of Tropical Medicine and Hygiene*, 82(2): 212–15.

Snow, R. W., Omumbo, J. A., Lowe, B., Molyneux, C. S., Obiero, J.-O., Palmer, A., *et al.* (1997) "Relation between severe malaria morbidity in children and level of Plasmodium falciparum transmission in Africa," *The Lancet*, 349(9066): 1650–4.

Soja, E. (2010) *Seeking Spatial Justice*. Minneapolis, MN: University of Minnesota Press.

Somerville, G. G., Diaz, S., Davis, S., Coleman, K. D., and Taveras, S. (2006) "Adapting the popular opinion leader intervention for Latino young migrant men who have sex with men," *AIDS Education and Prevention*, 18(Suppl. A): 137–48.

Spatial Networks (2014) *Fulcrum Mobile Location Leverage* [online]. Available at: <http://fulcrumapp.com/> (accessed 12 November 2014).

Srinivasan, V., Seto, K., Emerson, R., and Gorelick, S. (2013) "The impact of urbanization on water vulnerability: a coupled human–environment system approach for Chennai, India," *Global Environmental Change*, 23(1): 229–39.

St John, M., Durant, M., Campagna, P. D., Rehman, L. A., Thompson, A. M., Wadsworth, L. A., *et al.* (2008) "Overweight Nova Scotia children and youth: the roles of household income and adherence to Canada's food guide to healthy eating," *Canadian Journal of Public Health*, 99(4): 301–6.

Statistics Canada (2012) *Canadian Health Measures Survey: Cycle 2 Data Tables.* Ottawa: Statistics Canada.

Steckler, A. and Linnan, L. (eds) (2002) *Process Evaluation for Public Health Interventions and Research.* San Francisco: Jossey Bass.

Stokols, D. (1996) "Translating social ecological theory into guidelines for community health promotion," *American Journal of Health Promotion*, 10(4): 282–8.

Story, M., Kaphingst, K. M., Robinson-O'Brien, R., and Glanz, K. (2008) "Creating healthy food and eating environments: policy and environmental approaches," *Annual Review of Public Health*, 29: 253–72.

Story, M., Neumark-Sztainer, D., and French, S. (2002) "Individual and environmental influences on adolescent eating behaviors," *Journal of the American Dietetic Association*, 102(3): S40–S51.

Sturm, R. (2008) "Disparities in the food environment surrounding US middle and high schools," *Public Health*, 122(7): 681–90.

Sturm, R. and Cohen, D. (2009) "Zoning for health? The year-old ban on new fast-food restaurants in South LA," *Health Affairs*, 28(6): 1088–97.

Sturtz, S., Ligges, U., and Gelman, A. E. (2005) "R2WinBUGS: a package for running WinBUGS from R," *Journal of Statistical Software*, 12(3): 1–16.

Susser, M. (1995) "Editorial: The tribulations of trials – intervention in communities," *American Journal of Public Health*, 85(2): 156–8.

Swinburn, B., Egger, G., and Raza, F. (1999) "Dissecting obesogenic environments: the development and application of a framework for identifying and prioritizing environmental interventions for obesity," *Preventive Medicine*, 29(6): 563–70.

Swinburn, B. A., Sacks, G., Hall, K. D., McPherson, K., Finegood, D. T., Moodie, M. L., *et al.* (2011) "The global obesity pandemic: shaped by global drivers and local environments," *The Lancet*, 378(9793): 804–4.

Taylor, S. M., Ross, N. A., Cummings, K. M., Glasgow, R. E., Goldsmith, C. H., Zanna, M. P., *et al.* (1998a) "Community Intervention Trial for Smoking Cessation (COMMIT): changes in community attitudes toward cigarette smoking," *Health Education Research*, 13(1): 109–22.

Taylor, S. M., Ross, N. A., Goldsmith, C. H., Zanna, M. P., and Lock, M. (1998b) "Measuring attitudes towards smoking in the Community Intervention Trial for Smoking Cessation (COMMIT)," *Health Education Research*, 13(1): 123–32.

Taylor, S. M., Messina, J. P, Hand, C. C., Juliano, J. J., Muwonga, J., Tshefu, A. K., *et al.* (2011a) "Molecular malaria epidemiology: mapping and burden estimates for the Democratic Republic of the Congo, 2007," *PLoS One*, 6(1): e16420.

Taylor, S. M., van Eijk, A. M., Hand, C. C., Mwandagalirwa, K., Messina, J. P., Tshefu, A. K., *et al.* (2011b) "Quantification of the burden and consequences of pregnancy-associated malaria in the Democratic Republic of the Congo," *Journal of Infectious Diseases*, 204(11): 1762–71.

Tester, J. M., Stevens, S. A., Yen, I. H., and Laraia, B. A. (2010) "An analysis of public health policy and legal issues relevant to mobile food vending," *American Journal of Public Health*, 100(11): 2038–46.

Texas Department of State Health Services (2009) *Texas HIV Surveillance Report: 2009 Annual Report* [online]. Available at: <http://www.dshs.state.tx.us/hivstd/reports/default.shtm> (accessed 18 December 2010).

Texas Department of State Health Services (2011) *Texas HIV Surveillance Report: 2011 Annual Report* [pdf] Available at: <https://www.dshs.state.tx.us/hivstd/reports/HIVSurveillanceReport2011.pdf> (accessed 1 May 2015).

Texas Department of State Health Services (2013) *An Overview of HIV in Texas* [pdf]. Available at: <https://www.dshs.state.tx.us/hivstd/info/edmat/HIVAIDSinTexas.pdf> (accessed 1 May 2015).

Therrien, M. and Ramirez, R. R. (2000) *The Hispanic Population in the United States* [pdf]. Available at: <https://www.census.gov/prod/2001pubs/p20-535.pdf> (accessed 1 May 2015).

Thomas, R. E., McLellan, J., and Perera, R. (2013) "School-based programmes for preventing smoking," *Evidence-Based Child Health: A Cochrane Review Journal*, 8(5): 1616–2040.

Townshend, T. and Lake, A. A. (2009) "Obesogenic urban form: theory, policy and practice," *Health and Place*, 15: 909–16.

Tremblay, M. S., Colley, R. C., Saunders, T. J., Healy, G. N., and Owen, N. (2010) "Physiological and health implications of a sedentary lifestyle," *Applied Physiology, Nutrition, and Metabolism*, 35: 725–40.

Turner, B. (2010) "Vulnerability and resilience: coalescing or paralleling approaches for sustainability science?" *Global Environmental Change*, 20(4): 570–6.

Turner, B., Kasperson, R., Matson, P., McCarthy, J., Corell, R., Christensen, L., *et al.* (2003) "A framework for vulnerability analysis in sustainability science," *Proceedings of the National Academy of Sciences*, 100(14): 8074–9.

UNAIDS (1998) *HIV-related Opportunistic Diseases* [pdf]. Available at: <http://www.unaids.org/en/media/unaids/contentassets/dataimport/publications/irc-pub05/opportu_en.pdf> (accessed 20 February 2011).

UNEP (2011) *Environmental Assessment of Ogoniland*. Nairobi: United Nations Environment Program Report.

UNISDR (2004) *Living with Risk: A Global Review of Disaster Reduction Initiatives*. Geneva: United Nations International Strategy for Disaster Reduction.

US Department of Health and Human Services (1999) *Physical Activity and Health: A Report of the Surgeon General*. Atlanta, GA: US Department of Health and Human Services, Centers for Disease Control and Prevention, National Center for Chronic Disease Prevention and Health Promotion, The President's Council on Physical Fitness and Sports.

US Department of Health and Human Services (2011) *Annual Update of HHS Poverty Guidelines* [online]. Available at: <http://aspe.hhs.gov/poverty/11fedreg.shtml> (accessed 20 January 2012).

US Department of Health and Human Services (2013) *Stages of HIV Infection* [online]. Available at: <http://www.aids.gov/hiv-aids-basics/just-diagnosed-with-hiv-aids/hiv-in-your-body/stages-of-hiv/> (accessed 22 December 2014).

US Department of Health and Human Services (2014) *The Health Consequences of Smoking – 50 Years of Progress: A Report of the Surgeon General*. Atlanta, GA: US Department of Health and Human Services, Centers for Disease Control and Prevention, National Center for Chronic Disease Prevention and Health Promotion, Office on Smoking and Health.

van der Horst, K., Oenema, A., Ferreira, I., Wendel-Vos, W., Giskes, K., van Lenthe, F., *et al.* (2007) "A systematic review of environmental correlates of obesity-related dietary behaviors in youth," *Health Education Research*, 22(2): 203–26.

van der Horst, K., Timperio, A., Crawford, D., Roberts, R., Brug, J., and Oenema, A. (2008) "The school food environment: associations with adolescent soft drink and snack consumption," *American Journal of Preventive Medicine*, 35: 217–23.

van Hulst, M. and Yanow, D. (2016) "From policy 'frames' to 'framing' theorizing a more dynamic, political approach," *American Review of Public Administration*, 46(1): 92–112.

Vega, M. Y., Spieldenner, A. R., DeLeon, D., Nieto, B. X., and Stroman, C. A. (2010) "SOMOS: evaluation of an HIV prevention intervention for Latino gay men," *Health Education Research*, 26(3): 407–18.

Vine, M. M. and Elliott, S. J. (2014a) "Exploring the school nutrition policy environment in Canada using the ANGELO framework," *Health Promotion Practice*, 15(3): 331–9.

Vine, M. M. and Elliott, S. J. (2014b) "Examining local level factors shaping school nutrition policy implementation in Ontario, Canada," *Public Health Nutrition*, 17(6): 1290–8.

Vine, M. M., Elliott, S. J., and Raine, K. D. (2014) "Exploring implementation of the Ontario School Food and Beverage Policy at the secondary school level: a qualitative study," *Canadian Journal of Dietetic Practice and Research*, 75(3): 1–7.

Vos, T., Flaxman, A. D., Naghavi, M., Lozano, R., Michaud, C., Ezzati, M., *et al.* (2012) "Years lived with disability (YLDs) for 1160 sequelae of 289 diseases and injuries 1990–2010: a systematic analysis for the Global Burden of Disease Study 2010," *The Lancet*, 380(9859): 2163–96.

Vyas, S. and Kumaranayake, L. (2006) "Constructing socio-economic status indices: how to use principal components analysis," *Health Policy and Planning*, 21(6): 459–68.

Walensky, R. P., Paltiel, A. D., Losina, E., Mercincavage, L. M., Schackman, B. R., Sax, P. E., *et al.* (2006) "The survival benefits of AIDS treatment in the United States," *Journal of Infectious Diseases*, 194(1): 11–19.

Walker, B., Holling, C. S., Carpenter, S. R., and Kinzig, A. (2004) "Resilience, adaptability and transformability in social – ecological systems," *Ecology and Society*, 9(2): 5.

Wallace, R. G. (2003) "AIDS in the HAART era: New York's heterogeneous geography," *Social Science and Medicine*, 56(6): 1155–71.

Wallace, R. and Wallace, D. (1997) "Socioeconomic determinants of health: community marginalization and the diffusion of disease and disorder in the United States," *British Medical Journal*, 314(7090): 1341–5.

Wang, H., Qiu, F., and Swallow, B. (2014) "Can community gardens and farmers' markets relieve food desert problems? A study of Edmonton, Canada," *Applied Geography*, 55: 127–37.

Wang, X., Ouyang, Y., Liu, J., Zhu, M., Zhao, G., Bao, W., *et al.* (2014) "Fruit and vegetable consumption and mortality from all causes, cardiovascular disease, and cancer: systematic review and dose-response meta-analysis of prospective cohort studies," *British Medical Journal*, 349. DOI: 10.1136/bmj.g4490.

Watts, A. W., Mâsse, L. C., and Naylor, P.-J. (2014) "Changes to the school food and physical activity environment after guideline implementation in British Columbia, Canada," *International Journal of Behavioral Nutrition and Physical Activity*, 11: 50–9.

Watts, M. and Bohle, H. (1993) "The space of vulnerability: the causal structure of hunger and famine," *Progress in Human Geography*, 17(1): 43–67.

Williams, T. N., Mwangi, T. W., Wambua, S., Alexander, N. D., Kortok, M., Snow, R. W., *et al.* (2005) "Sickle cell trait and the risk of Plasmodium falciparum malaria and other childhood diseases," *Journal of Infectious Diseases*, 192(1): 178–86.

Willis, C. D., Riley, B. L., and Jolin, M. A. (2013) *Improving the Performance of Partnerships for Chronic Disease Prevention: Report on a Planning Meeting*. Waterloo, Canada: Propel Centre for Population Health Impact, University of Waterloo.

Willis, C. D., Riley, B., Best, A., and Ongolo-Zogo, P. (2012) "Strengthening health systems through networks: the need for measurement and feedback," *Health Policy and Planning*, 27(Suppl. 4): 62–6.

Wills, W., Grunert, K. D., Fernández Celemín, L., and Bonsmann, S. S. G. (2009) "Do European consumers use nutrition labels?" *AgroFOOD Industry Hi-Tech*, 20(5): 60–2.

Wilson, P. A. and Yoshikawa, H. (2004) "Experiences of and responses to social discrimination among Asian and Pacific Islander gay men: their relationship to HIV risk," *AIDS Education and Prevention*, 16(1): 68–83.

Wilton, L., Herbst, J. H., Coury-Doniger, P., Painter, T. M., English, G., Alvarez, M. E., et al. (2009) "Efficacy of an HIV/STI prevention intervention for black men who have sex with men: findings from the Many Men, Many Voices (3MV) project," *AIDS and Behavior*, 13(3): 532–44.

Wise, R., Fazey, I., Stafford Smith, M., Park, S., Eakin, H., Archer Van Garderen, E., et al. (2014) "Reconceptualising adaptation to climate change as part of pathways of change and response," *Global Environmental Change*, 28: 325–36.

Woodland, R. H. and Hutton, M. S. (2012) "Evaluating organizational collaborations: suggested entry points and strategies," *American Journal of Evaluation*, 33(3): 366–83.

World Health Organization (1986) *The Ottawa Charter for Health Promotion* [online]. Available at: <www.phac-aspc.gc.ca/ph-sp/docs/charter-chartre/index-eng.php> (accessed 28 November 2015).

World Health Organization (2000) *WHO Expert Committee on Malaria: Twentieth Report* [pdf]. Available at: <http://apps.who.int/iris/bitstream/10665/42247/1/WHO_TRS_892.pdf> (accessed 27 November 2015).

World Health Organization (2009) "Seat-belt celebrates its 50th birthday" [online]. Available at: <http://www.who.int/violence_injury_prevention/media/news/2009/13_08_09/en> (accessed 3 April 2016).

World Health Organization (2012a) *Global Strategy for Dengue Prevention and Control 2012–2020*. Geneva: World Health Organization.

World Health Organization (2012b) *Controlling the Global Obesity Epidemic*. Geneva: World Health Organization.

World Health Organization (2014) *World Malaria Report 2013* [online]. Available at: <http://www.who.int/malaria/publications/world_malaria_report_2013/en/> (accessed 27 November 2015).

World Health Organization and Food and Agricultural Organization of the United Nations (2004) *Fruit and Vegetables for Health: Report of a Joint FAO/WHO Workshop, 1–3 September 2004, Kobe, Japan*. Geneva and Rome: WHO and FAO.

Woulfe, J., Oliver, T. R., Siemering, K.Q., and Zahner, S. (2010) "Multisector partnerships in population health improvement," *Preventing Chronic Disease*, 7(6): A119.

Wrigley, N. (2002) "'Food deserts' in British cities: policy context and research priorities," *Urban Studies*, 39(11): 2029–40.

Young, T. (2005) *Population Health*. New York: Oxford University Press.

Zablotska, I., Holt, M., and Prestage, G. (2011) "Changes in gay men's participation in gay community life: implications for HIV surveillance and research," *AIDS and Behavior*, 16: 666–75.

Index

Milton Keynes UK
Ingram Content Group UK Ltd.
UKHW040100071024
449327UK00019B/699